WA 1241498 0

D1327923

GIS in Law Enforcement

Essential Reading

Also available from Taylor & Francis

Mapping and Analysing Crime Data
Lessons from research and practice
Edited by Alex Hirschfield and Kate Bowers
both at University of Liverpool, UK
ISBN 0–7484–0922–X (hardback)

Manual of Geospatial Science and Technology
Edited by John Bossler
Ohio State University, USA
ISBN 0-7484-0924-6 (hardback)

Community Participation and Geographic Information Systems
Edited by W. J. Craig, T. M. Harris and D. Weiner
ISBN 0-415-23752-1 (hardback)

Web Cartography
Edited by M. J. Kraak and A. Brown
ITC, Enschede, Netherlands
ISBN 0-7484-0868-1 (hardback)
ISBN 0-7484-0869-X (paperback)

GIS, Organisations & People
A socio-technical approach
J. Petch and D. Reeve
ISBN 0-7484-0270-5 (paperback)

Information and ordering details

For price availability and ordering visit our website: www.gisarena.com
Alternatively our books are available from all good bookshops.

GIS in Law Enforcement

Implementation issues and case studies

Edited by
Mark R. Leipnik and
Donald P. Albert

LONDON AND NEW YORK

First published 2003
by Taylor & Francis
11 New Fetter Lane, London EC4P 4EE

Simultaneously published in the USA and Canada
by Taylor & Francis Inc
29 West 35th Street, New York, NY 10001

Taylor & Francis is an imprint of the Taylor & Francis Group

Typeset in 10/12 Sabon by
Newgen Imaging Systems (P) Ltd, Chennai, India
Printed and bound in Great Britain by
TJ International Ltd, Padstow, Cornwall

British Library Cataloguing in Publication Data
A catalogue record for this book is available from the British Library

Library of Congress Cataloging in Publication Data
A catalog record for this book has been requested

ISBN 0-415-28610-7

To

Joanne and Michelle
&
Elizabeth, Kenny, and Julie

Contents

Contributors

Donald P. Albert, PhD, Assistant Professor, Department of Geography and Geology, Sam Houston State University, Huntsville, Texas.

Laurie Anderson, Planner/Analyst, Spokane Police Department, Spokane, Washington.

Billy Asbell, Lt. Colonel Deputy Chief, United States National Guard Bureau, Special Projects Division, Washington, DC.

Gary Birchall, Criminal Intelligence Analyst, South Yorkshire Police Authority, Sheffield, England.

John Bottelli, GIS Specialist, Spokane County Government, Spokane, Washington.

Tim Burns, Justice Information Analyst, Department of Justice Coordination, Pinellas County, Florida, Clearwater, Florida.

Tom Casady, Cheif Lincoln Police Department, Lincoln, Nebraska.

Kenneth Clontz, PhD, Department of Law Enforcement and Justice Administration, Western Illinois University, Macomb, Illinois.

Tony Cooper, GIS Analyst, King County Government, Seattle, Washington (Formerly GIS Specialist, Spokane County Government, Spokane, Washington).

Andrew Costello, PhD, Research Fellow, South Yorkshire Police Authority, Sheffield, England.

Tom Evans, Crime Analyst, Pinellas County Sheriff's Department, Clearwater, Florida.

Bryan Hill, Crime/Traffic Analyst, Phoenix Police Research and Analysis Unit, Phoenix, Arizona.

Robert Hubbs, Crime Analysis Unit, Knoxville, Police Department, Knoxville, Tennessee.

Dennis Kidwell, Criminal Investigation Division, Waco Police Department, Waco, Texas.

Mark R. Leipnik, PhD, Associate Professor, Department of Geography and Geology; Director of GIS Laboratory, Texas Research Institute for Environmental Studies, Sam Houston State University, Huntsville, Texas. *Point of contact for questions pertaining to the book*. Phone: (936) 294-3698; Fax: (936) 294-3940; E-mail: geo_mrl@shsu.edu.

Jamie May, Crime Analyst, Crime Analysis Unit, Overland Park Police Department, Overland Park, Kansas.

Albert Mellis, Cheif, Waco Police Department, Waco, Texas.

J. Gayle Mericle, PhD, Department of Law Enforcement and Justice Administration, Western Illinois University, Macomb, Illinois.

Joseph Messina, PhD, Assistant Professor, Department of Geography, Michigan State University, East Lansing, Michigan.

Andreas Olligschlaeger, PhD, TruNorth Data Systems, Inc. 1260 Freedom Crider Rd., Freedom, Pennsylvania.

Derek J. Paulsen, PhD, Assistant Professor, Department of Political Science and Criminal Justice, Appalachian State University, Boone, North Carolina.

Erika D. Poulsen, PhD, Candidate, Department of Geography, Rutgers University, Newark, New Jersey.

Kristin Preston, GIS Programmer/Analyst, Pinellas County Sheriff's Office, Clearwater, Florida.

D. Kim Rossmo, PhD, Director of Research, The Police Foundation. Washington, DC.

Ariane Schmidt, Public Safety GIS Administrator/Planner, Spokane Fire Department (Formerly Planner/Analyst, Spokane Police Department) Spokane, Washington.

Gerry Tallman, Manager, Crime Analysis Unit, Overland Park Police Department, Overland Park, Kansas.

Ian Von Essen, GIS Network Manager, Spokane County Government, Spokane, Washington.

Carl Walter, Crime Analyst, Boston Police Department, Boston, Massachusetts.

Julie Wartell, PhD, Senior Research and Technology Associate, Institute for Law and Justice, San Diego, California.

Foreword

Maps, technology, and the search for treasure

In Robert Louis Stevenson's 1882 adventure story, *Treasure Island*, Jim Hawkins discovers a map at the bottom of a pirate's sea-chest. Three crosses in red ink on the chart show how to find the location of Captain Flint's buried treasure of gold bars, doubloons, and guineas. Hawkins's discovery leads to adventure, a sea voyage, mutiny, and finally murder as blood is spilled over the map's possession. What was it that made a piece of paper so valuable?

The value of the map came from its information content – the locations of the crosses that marked the buried pirate treasure. All maps are a display of information, simple or complex, public or secret. They allow us to place ourselves within the overall context of the world; they help us know where we are, where we want to go, and how to get there. Maps have been indispensable since the early days of human exploration, and in fact, have been interactive with progress. In a cyclical process, exploration leads to discovery, which results in more complete and accurate maps, which prompt yet further exploration and innovation.

Representations of geography have been realized in different ways, at different times, around the world. The earliest maps used symbols to orient the traveller. Polynesians wove maps of reeds studded with seashells to show the direction of ocean currents and position of islands. Modern maps, the children of Cartesian coordinate systems and trigonometry, reflect the state of our technology and our need for aid in precise global navigation and local and regional design and engineering. Cartography is the product of cultural, social, and economic influences upon an interpretation of the physical environment. There are politics and religion in map making. There are gender differences. And there can be deceit created by the choice of what to show and how, and what not to. Most of all, maps are a function of our need for spatial information.

Technology significantly enhances the information value of maps and consequently their importance. The lodestone and compass oriented travelers with direction. The astrolabe and chronometer allowed ships to establish their latitude and longitude and opened up the New World. Today, global positioning systems (GPS) accurately determine the location of remote places to within a few meters. And geographic information systems (GIS) are revolutionizing our use of spatial information.

Geographic information systems are simply computerized maps linked to descriptive information. They have been described as two-dimensional databases (though it should be noted geographical analysis can involve up to four dimensions). Geography is ubiquitous. We live in it. We move in it. It surrounds us and our activities, whether we are engaged in a trip to the corner store or a journey to Mount Kilimanjaro. We need to understand this geography in our daily lives and in our decision-making processes. Here lies the power of geographic information systems. Data on a diverse array of issues can be introduced and integrated through the common denominator of location. And these displays of interrelationships allow us to extract information from the data.

Geographic information systems supercharge the crime mapping process. They provide the ability to do more things, faster and more precisely, with a potential that was not previously possible. GIS has the ability to make crime mapping part of a police department's organizational routine. Ease of use, price, versatility, speed, and functionality equate to feasibility. But GIS is only part of the equation. Computer-aided dispatch (CAD) and records management systems (RMS), powerful desktop computing, and color printers connect in a more or less integrated technology stream. CompStat (which is short for comparative – not computer – statistics), first implemented by the New York Police Department in 1994, provides a good example of this potential. Combining crime analysis through mapping, problem solving, and managerial accountability, CompStat and its descendents have integrated GIS use into policing in a powerful manner.

Who are the customers for crime mapping and analysis within police organizations? The chapters in this edited book cover a range of GIS-related topics, including implementation issues, data sharing and database design, small town and statewide systems, community problems, counterdrug work, criminal investigations, aerial photography, the journey to crime, and future directions. This breadth of coverage underlines the many appropriate roles of GIS in law enforcement. It is a specific tool – a means, not an end – to various and diverse objectives. The objectives are tied together in the relevance of spatial information and its management to their accomplishment. Therefore, the users of GIS-related technologies include police leaders, managers, supervisors, investigators, and patrol officers. This means that communication and product credibility at all organizational levels are important concerns for the growth of GIS-based crime mapping.

But, as powerful as this technology is, there are clear limitations. Word processing is not equivalent to literature. We must be wary of "fast food" crime analysis, where just a few key strokes can run a software routine, producing what has been described as "pretty and meaningless" maps. It has been said that the introduction of new technology has three stages. First, we use it to do what we did before, in the same manner, just faster. Second, we use it to find new and better methods to do what we did before. And third, we find new, previously inconceivable, tasks to accomplish.

So how do we go beyond automated pin mapping to realize the full potential of geographic information systems in policing? It is an understanding of theory that helps us develop the knowledge necessary to effectively guide police actions. If we want our results to be significant, we should heed the warnings of John Eck and others regarding the risks of atheoretical crime mapping. Just as important as technological developments, have been the theoretical advances in environmental criminology that provide a context for understanding the geography of crime. This in turn guides spatial analysis, the knowledge component of crime mapping. David Weisburd has pointed to the study of crime and place, which has provided a better understanding of both aggregate and individual criminal spatial behavior, as an important contributor to the growth and utility of computerized crime mapping. Research is how we test and improve our theories. And experience provides the reality check.

Appropriately guided by theory, research, and experience, where can GIS take us? While there are many possibilities, I suggest some of the more important areas include the following. The accurate prediction of future crime events is the Holy Grail of crime analysis, and this will continue to be a tremendous challenge, both technically and theoretically. High-rise development is common in the urban built environment, and vertical analyses such as those being conducted by George Rengert will help us understand crime in three-dimensional space. A more complete understanding of the fourth dimension will allow us to analyze, in an integrated fashion, rhythms and cycles in both time and space, even incorporating the influences of weather and lighting patterns. And to be effective, police agencies must grasp the dynamics of displacement and diffusion, spatial, temporal, and otherwise.

Geographic information systems are the compasses of crime mapping. They turn theoretical potential into organizational reality. But there's no buried gold at the end of this search. Rather, the treasure is information and knowledge; through them we can achieve better policing and safer communities.

D. Kim Rossmo
Police Foundation, Washington, DC.

Acknowledgements

The authors wish to specifically recognize all of the unsung police officers, GIS analysts, technicians and other staff employees of the dozens of law enforcement agencies, universities and firms that contributed material included as chapters, as focus boxes and as maps and images used in figures throughout the book. For every listed author, there were often scores of staff members that made significant contributions; we cannot list them all, nor do we know the names and titles of most of these persons. However, we thank them all for their continuing efforts made in the public interest and for their specific contributions to this book. However, we would like to single out for special thanks Melinda Higgins of Georgia Tech University, Elayne Starkey of the Government of the State of Delaware, Milan Mueller of The Omega Group in San Diego, California and Danny Santos of the Austin, Texas Police Department, all of whom provided assistance of particular value.

In addition, the editors wish to recognize Dr Roy B. Leipnik, Professor of Mathematics at the University of California at Santa Barbara for proof-reading the entire manuscript and Dr C. Allen Williams, Chair of the Department of Geography and Geology at Sam Houston State University for reviewing parts of the book. Barbara Jones of the Academic Enrichment Center at Sam Houston State University also provided editorial assistance on several chapters. We wish to acknowledge the International Association of Police Chiefs for their permission to adapt an article appearing in *The Police Chief* (September 2000, pp. 34–49) for inclusion in this volume. We also wish to thank the editors of *Geospatial Solutions Magazine* for kindly allowing reuse of material featured in an article entitled "*Coordinates of a Killer*" which appeared in the November, 2001 issue and which contained a condensed version of the case presented in full in Chapter 12. The Office of Research and Sponsored Programs at Sam Houston State University provided partial funding for this project in the form of several Faculty Research Enhancement Grants. To all these persons and sources of support we express our gratitude.

Screen shots using Microsoft Windows Explorer are reprinted by permission from Microsoft Corporation. In this context, Windows and Explorer are trademarks of Microsoft.

ArcView Graphical User Interface is the intellectual property of ESRI and is used herein with permission. Copyright © 1996–2001 ESRI. All rights reserved.

Screen shots of rendered maps © 2002 MapInfo Corporation. All rights reserved. MapInfo® is a registered trademark of MapInfo Corporation.

CrimeView® related images are used with kind permission of Milan Müller, President, The Omega Group Inc., San Diego, CA.

CrimeStat is copyrighted by and the property of Ned Levine and Associates. The name CrimeStat is a registered trademark of Ned Levine and Associates. crimestat@nedlevine.com

Ned Levine, *CrimeStat*: A Spatial Statistics Program for the Analysis of Crime Incident Locations (v. 1.1). Ned Levine & Associates, Annandale, VA, and the National Institute of Justice, Washington, DC, July 2000.

Part I

Implementation issues

This book is primarily intended to help potential geographic information system (GIS) users. Current GIS practitioners and managers in law enforcement get the most out of what geographic information systems have to offer. In order to realize the benefits of GIS, an often complex implementation process must be successfully negotiated. The first part of this book discusses the vitally important issue of implementation. For most police agencies considering the adoption of GIS for use in crime mapping and analysis, implementation represents the major stumbling block. It will be an issue that continues to require significant effort as the technology matures and becomes more widely disseminated. This book provides managers and practitioners in police agencies a wide range of perspectives on implementation issues including a detailed step-by-step overview of the implementation process, answers to common implementation questions, and advice on major implementation issues. More focused chapters deal with implementation in small and medium-sized departments, database design issues, methods for sharing and disseminating data and on future directions for GIS in law enforcement. The issues raised in these chapters are highlighted by focus boxes that provide specific examples of police agencies that have grappled with such issues as geo-coding or posting of crime maps on the Internet and successfully integrated GIS into the department's workflow to become an integral part of daily operations. Materials presented in this book are drawn from throughout the US, and around the World. Figure I on the following page illustrates the locations of cases and materials from the US presented in the text.

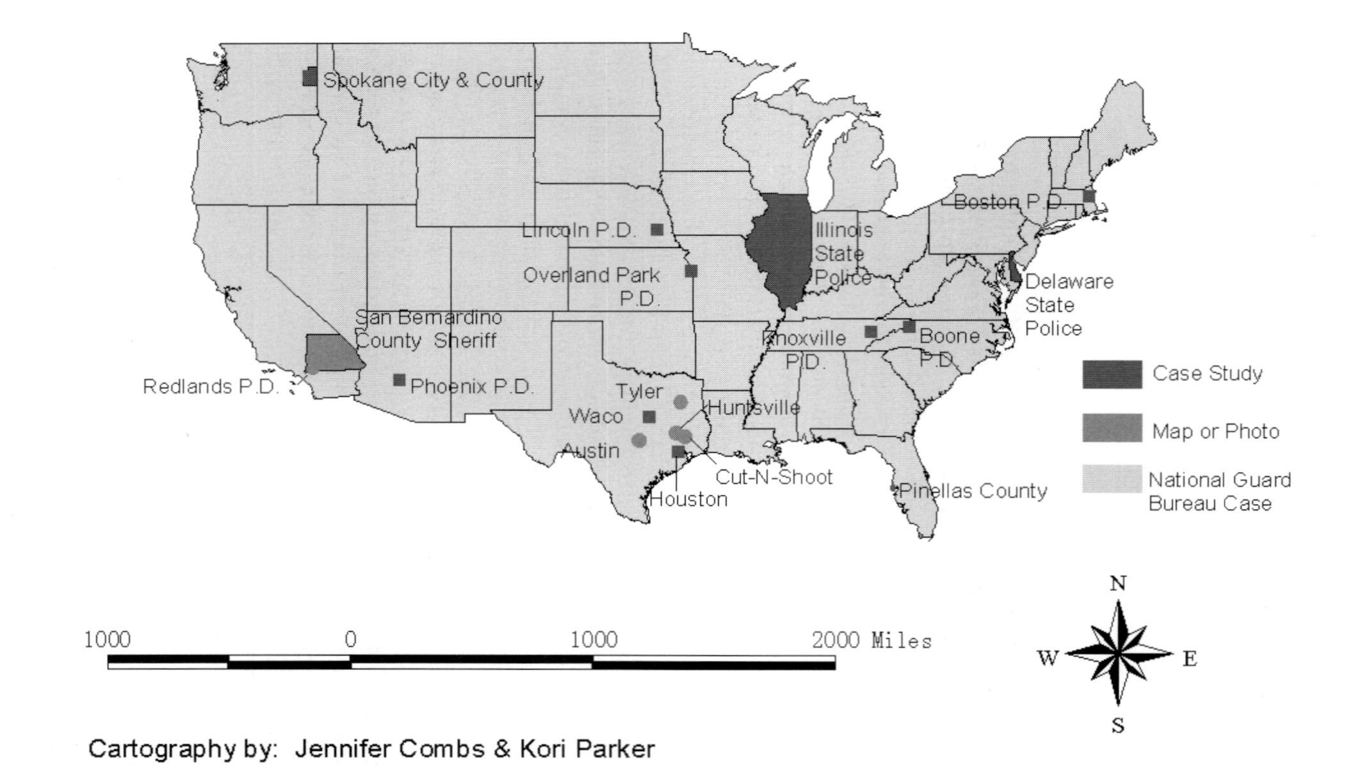

Figure I.A Distribution in the United States of America of cases, focus boxes and maps and other images presented in the book are shown on this map. Note that examples from many regions are presented as GIS technology is widely used in Law Enforcement in the US today. Note also that material from South Yorkshire, England; Abbotsford, British Columbia, Canada; and Iceland is also presented.

1 How law enforcement agencies can make geographic information technologies work for them

Mark R. Leipnik and Donald P. Albert

A geographic information system (GIS) is a powerful technological tool for municipal police departments and other law enforcement agencies. Typical GIS users in law enforcement include crime analysts, computerized crime records management personnel, police executives, shift supervisors, patrol sergeants, and even patrol officers. All current and potential users of GIS can benefit from a better understanding of what GIS is and what makes GIS special and different from other information technologies such as databases, computer-aided dispatching, and computer aided design (Leipnik and Albert 2000). This chapter reviews the basic concepts and applications of GIS in law enforcement activities.

What is GIS?

GIS is an abbreviation for geographic information systems. Defining GIS is often elusive because there is no universal definition; however, there is general agreement that GIS is a powerful tool for spatial analysis. It captures, stores, manipulates, analyzes, displays and queries geographic data and generates cartographic and statistical outputs.

Implementing a GIS requires people (officials, managers, technicians), data, software, hardware, and procedures. Overarching these components should be the application of sound geographic reasoning that involves asking geographic questions, acquiring, organizing and analyzing geographic information, and answering geographic questions. Therefore, geographic reasoning and GIS provide crime analysts an effective one-two punch in maximizing the potential benefits of crime mapping.

GIS supports various spatial analysis functions.

1. View and query of, classification of, and measurement of spatial data.
2. Overlay operations that involve adding, removing or reordering map layers.
3. Neighborhood operations that select features contained within buffer zones generated around points, lines, and areas.
4. Network analysis functions such as shortest route determination.

Space limitations prevent further elaboration; however, readers might examine *Geographic Information Systems: A Management Perspective* by Stan Aronoff (1989) for a more detailed discussion of GIS functions.

Characteristics of law enforcement GIS

Examples of spatial data relevant to law enforcement include point (crime locations), line (streets), and area (precinct boundaries) features. These geographic features are often separated into overlapping thematic layers (schools, parks, political boundaries, criminal offenses, traffic, and 911 calls, to name a few examples). GIS provides users the flexibility to combine or separate data into as many or as few layers as needed. All layers in a GIS have a common coordinate system, frequently either Universal Transverse Mercator (UTM) or State Plane Coordinates in the US. A GIS can manage multiple-coordinate systems and projections and support cartographic transformations of the data so that data can be imported in another format and included into the GIS.

A GIS can also usually accept data that is in a raster format including scanned images of maps, digital aerial photography, or remotely sensed imagery. Although a GIS can bring this data in as a new layer, it will not have the same information content as existing data stored in a vector format. Examples of typical vector-based data sets are the street centerline data most departments use to serve as a base-map for crime mapping.

Tables are used to store attribute data in a GIS. These tables have a series of records that are linked to the map features, such as points representing crime incidents, lines representing street segments, or polygons representing precincts. This attribute data can be extracted, revised, and managed separately from the spatial data and, in many cases, is of greater importance than the spatial data. A common approach in law enforcement is to collect the attribute data from crime incident reports entered into a stand-alone database and then selectively extract incident data for insertion into the GIS. Often, this requires geo-coding of the location of an incident. Usually, the street address of an incident is compared to a database linked to the GIS that contains street names and address ranges for the right and left sides of every city block. The creation of this data is a challenge for a police department; fortunately, the US Census Bureau has, since 1990, created just such a data set. It is called TIGER (topologically integrated geographic encoding and referencing) and is available to the public nationwide. This TIGER street data serves as the basis for most crime incident geo-coding and hence for most crime mapping in the US at present. More recently, global positioning systems (GPS) have become a viable alternative for determining the location of incidents for inclusion in a GIS. This is particularly true of incidents that occur where street addresses are unavailable or do not tie the incident to a limited area.

When a new incident is entered into a GIS, a point is inserted at the corresponding location, a database table is opened, and a new record is inserted into this table. In general, a special symbol is used to denote each class of incident. Thus, serious ("part-one") incidents might be subdivided into burglaries, robberies, and assaults with a unique symbol associated with each. Some departments choose to use connotative symbols. For example, symbols in the shape of a gun, knife, cracked house, or tombstone might be used to represent, the location of, respectively, a robbery, assault, burglary, and homicide. Most departments choose to use geometric figures (squares, stars, circles) of various colors to denote different types of crimes or crimes occurring at differing time periods.

Aiding crime analysis

The most common approach is to have a crime analyst use GIS to analyze the spatial and temporal factors associated with a series of crimes or to detect patterns, trends, and exceptions. In most police departments, crime analysts view GIS as an important but non-essential tool. They may generate reports on only selected crimes or create crime maps occasionally in response to particular types of incidents such as serial robberies or rapes. The reason that many departments use GIS selectively is that it is a relatively new technology in policing. Furthermore, geo-coding all crime incidents or even all major crimes is a laborious task, and in many departments adequate resources have not been provided to support maintenance of a GIS. Thus, crime analysts are often assigned the additional task of crime mapping, using a simple GIS program and a less-than-perfect street data set from TIGER with which to work. This is not to say that the crime maps produced periodically are not useful, or that crime mapping of specific incidents or classes of incidents is not valuable. Examples of serial rapists, murderers, and robbers caught through the use of deployment and response strategies – analyzed and generated with a GIS – exist for many cities. The United States Federal Bureau of Investigation occasionally uses GIS; the Royal Canadian Mounted Police (RCMP) and German Federal Police (BKA) do so also.

In addition to helping to solve crimes associated with a particular perpetrator or gang, many crime analysts use GIS to examine crime patterns associated with a particular locality, say, a liquor store or a motel that may experience a significant number of crimes in its vicinity. Sometimes, it is difficult to identify locations that for whatever reason seem to act as crime magnets. Frequently, the incidents may not actually be on the premises of a liquor store or motel, but crimes such as assaults, prostitution, and drug dealing may be concentrated in close proximity to the offending establishment.

A function standard to all GIS programs, "buffer zone generation," allows an analyst to generate a zone (often a circle with a specified radius) centered on the location of interest and then extract all incidents within that zone. These incidents will appear in a different color on the screen, and the corresponding records from the database can automatically be selected from the entire mass of records. For example, all incidents within 500 feet of a particular liquor store can be easily identified. The San Diego, California and Charlotte-Mecklenburg, North Carolina police departments use this approach to identify motels and liquor stores that have an unusually large number of crime incidents in close proximity. If it is established that there is a link between a motel and prostitution and drug dealing, or a liquor store with assaults, vagrancy, and petty theft, enforcement actions are directed at the establishment in question. These actions are not necessarily geared toward proving criminal activity on the part of the business. Rather, civil enforcement actions or removal of liquor licenses are the most common responses. In some cases, the police department cooperates with city building inspectors, fire marshals, planning officers, and district attorneys to rid the community of these magnets for crime. GIS has not only helped identify these establishments, but also has been used in evidence in court and liquor control board proceedings as graphic (and geographic) proof of the localized crime, all too frequently associated with a small but distinct minority of businesses in the community.

Improving command-level decision making

Command-level executives within police agencies are now using GIS to help decide how to deploy resources and where to locate facilities. Perhaps most visibly, the New York City Police Department has made GIS an essential component of the COMSTAT (comparative statistics) process. The crime statistics considered on a monthly (or more frequent) basis by decision makers in order to better deploy resources and target specific problems are displayed and analyzed using a GIS. The GIS helps decision makers see crime patterns graphically in relation to community features and precinct boundaries. In particular, clusters of incidents become very visible.

Once decision makers recognize problems, either in terms of spatial or temporal "hot spots" of crime, the COMSTAT process focuses resources on combating these problems. If displacement of crime results, the next monthly meeting can observe that trend and make necessary adjustments in deployment. In particular, the use of GIS in this process has helped the department identify crime hot spots that cut across precinct boundaries. Before the advent of COMSTAT, for example, an area of street crime associated with crack cocaine dealing near the convergence of three precincts was likely to go unnoticed by all three precincts. GIS can help the involved precincts to recognize the problem by displaying both crime incident locations and precinct boundaries. Then, resources within and among precincts can be re-allocated appropriately.

GIS also helps the New York City Police Department perform spatial and temporal analysis simultaneously. In this way, incidents occurring at a particular time of day, or on weekends, will be displayed on a crime map. Thus, the technology will highlight an area with high rates of street crime on Friday and Saturday nights, for instance, or an area with daytime mid-week residential burglaries. The relevant commanders can then make changes in scheduling and, if necessary, approve overtime. Many departments are using GIS to assist deployment, though few use it as systematically as the New York City Police Department.

Authorities can use GIS to determine the best location for new facilities, as well. Numerous fire departments and other agencies involved in emergency response have used GIS to locate facilities. New York City uses it's GIS to plan and locate facilities and also to respond to emergencies such as the attack on the world trade center, in which GIS played a key role in response and recovery efforts. As GIS becomes more prevalent and accepted in law enforcement, police departments will find it a useful tool to locate community outreach "storefront" offices, new substations, and, ultimately, new headquarters and other facilities and to coordinate emergency response and security related activities.

Assisting patrol and community outreach activities

GIS can be a great help to individual patrol officers and officers working in community-oriented policing programs. A number of departments routinely generate crime maps attached to tabular lists of specific incidents for a given period (frequently a month or week). These bulletins are copied and distributed at briefings or even at the beginning of every shift. In several departments that provide a computer (desktop or laptop) to most officers, an effort is underway to use internal local area networks to make this data available digitally. Several other police departments, including those in Chicago and Houston have copies of a GIS program in every precinct or district office. The effort to disseminate the results of crime incident mapping and spatial analysis of crime down to the level of the patrol officer is now underway. Since a GIS can focus on the beat, precinct, or community, the maps can focus on issues pertinent to particular patrol officers. Creating incident maps for most possible types of crimes on a routine basis is a major effort. In particular, getting the incident reports into a GIS in a timely and accurate manner deserves special emphasis.

An example of a department that has done an outstanding job of disseminating information down to the patrol level is the Chicago Police Department. This department puts many categories of crime incidents into its GIS on a daily basis and has achieved a remarkable 99.8 percent success rate in geo-coding these incidents by systematically making sure that addresses recorded in crime reports are consistent with the street names and address ranges stored in the GIS. This involves carefully checking the street map in the GIS for accuracy, and removing glitches such as missing streets, or streets with out-of-date names or address ranges. More importantly, Chicago has impressed on its officers the need to record the street addresses of incidents accurately and make an extra effort to spell street names correctly. This effort is being rewarded with crime incident bulletins and on-line crime incident maps delivered into the hands of patrol officers and others in the chain of command that are current as of the preceding day and, in some cases, the preceding shift.

Conclusion

GIS is a technology that is accessible to any police department with the resources to assign a crime analyst, information specialist, or staff member reasonably competent in the use of computers to the project for a portion of his or her workday. GIS offers law enforcement professionals throughout the department a powerful tool to see and understand spatial and temporal patterns of crime and police responses to these issues in a profoundly new way. GIS applications range from the simple to the highly sophisticated, but the hundreds of cities and many county law enforcement agencies currently using GIS started by gaining an initial understanding of this exciting technology.

References

Aronoff, S. (1989) *Geographic Information Systems: A Management Perspective*. Ottawa, Canada: WDL Publications.

Leipnik, M. and D. Albert (2000) How law enforcement agencies can make geographic information technologies work for them. *The Police Chief* 67: 34–49.

2 Overview of implementation issues

Mark R. Leipnik and Donald P. Albert

Implementation of a geographic information system within a law enforcement agency has many issues in common with developing a GIS for any municipal (or other level) governmental agency. This is fortunate since agencies throughout the world have been struggling, and generally succeeding, in implementing GIS for decades (Leipnik *et al.* 1993). Many of the obstacles likely to be encountered in the context of GIS implementation in law enforcement have been overcome by other agencies. Methods, strategies, cadres of consultants, and a myriad of training and education materials are available to all governmental entities to ease the implementation process. Viewing implementation as a structured multi-step process is perhaps the most important avenue by which implementation can be accomplished in a rapid and cost-effective manner. Merely muddling through or relying exclusively on the advice of a generally self-interested salesperson from a GIS vendor, is at least a recipe for added expense and/or unnecessary delays and false starts, if not a recipe for disaster. This chapter outlines a simplified step-by-step method for making implementation of GIS an efficient and intelligent process (Korte 1992).

This chapter also addresses many questions and issues that law enforcement agency management level personnel and potential GIS users within the organization may have. Some of these issues are unique to law enforcement, while others are not only of special importance to law enforcement but also occur in other application areas. Some issues are ubiquitous in all information systems technology, such as the need for basic education, ongoing training, and continual maintenance and upgrading of information technology hardware and software.

The GIS implementation process

A systematic process for implementation of a GIS is an ideal situation rarely achieved in any organization. Yet, it is valuable to recount the steps involved in a rational and systematic implementation process, and where possible make an effort to follow some, if not all, of the steps in such an "ideal" process. Figure 2.1 provides a schematic diagram of the steps in an

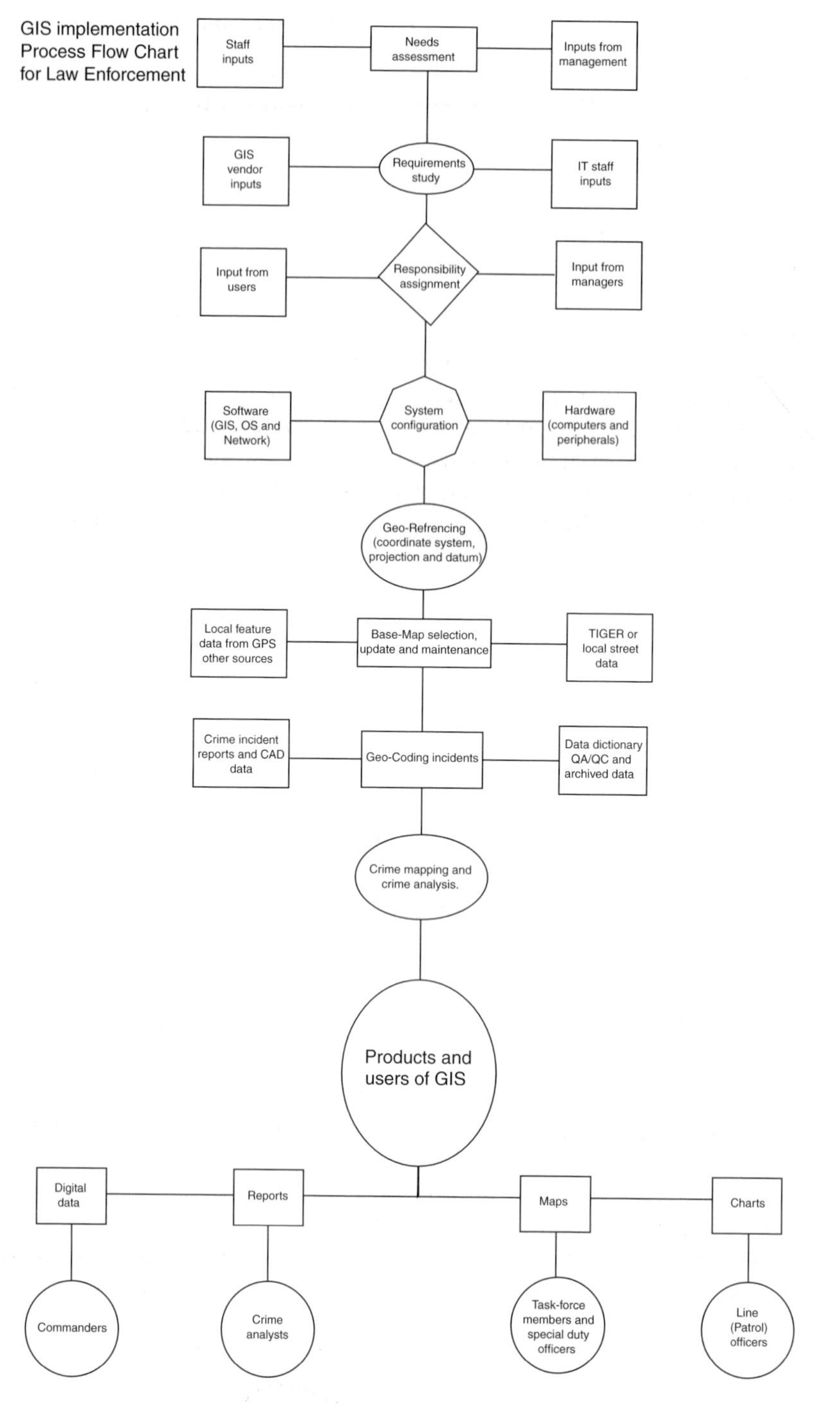

Figure 2.1 Implementation flow chart shows the numerous steps in a rigorous implementation strategy. *Flowchart prepared using adobe illustrator by Mark Leipnik.*

ideal process. This process would begin with identification of the geospatial data management and analysis needs within an organization, say a police department.

Needs assessment

A *needs assessment* could simply be an evaluation of needs by a relevant individual such as a crime analyst, information systems manager, or operations manager. Alternatively, a more formal process such as a survey of potential users or even a study undertaken by an outside consultant might be warranted, particularly in a large department. At any rate, this *needs assessment* should focus on what tasks a GIS could be applied to and who in the organization would likely use the GIS. This would provide the decision maker with some useful insights into how and by whom a GIS would be employed. Typically, a needs assessment would consider four functional areas where GIS could be helpful.

Firstly, at the *command level*, the need to support decision-making about deployment of officers could be supported with monthly maps and statistical reports of the spatial and temporal distribution of crime incidents by district (precinct). Thus, the spatial and temporal distribution of incidents by type for all major (part 1) crimes might be generated for the month of August for the 72nd precinct of New York City, and then viewed in conjunction with regular time and overtime hours for line and task force officers in that precinct. In order for such a precinct-by-precinct analysis to be really useful it would need to take into account the possibility that crime might be concentrated near the boundary of a given precinct (McGuire 1998). Thus, the unwritten law of cartography that "anything interesting occurs at the corner of four map sheets" has its counterpart in crime mapping since a "hot spot" for crime may spill over into adjacent areas of three or four precincts. If the deployment of resources is viewed for only one precinct in isolation, such "border" incidents will tend to be unrecognized by all of the involved precincts. Command level decision makers have to be vigilant for this issue, and a GIS can help identify this potential problem.

Secondly, a *crime analyst* might need access to his or her own copy (or access to a copy through a site license) of a GIS to create specialized crime density and pattern analyses for investigation of a specific series or types of crime. Thus, a crime analyst might want a map portraying locations of burglaries of homes that backed up to bicycle or walking paths, for example, where entry was made through a back door in The Woodlands, Texas or locations of pawn shops and jewelry stores vulnerable to a gang of masked shotgun wielding robbers in the greater Baltimore, Maryland area, or the spatial and temporal distribution of sexual assaults suspected of being perpetrated by a serial rapist in Las Vegas, Nevada or Detroit, Michigan. See Figure 2.2 for a crime-analyst generated map of burglaries from automobiles and Figure 2.3 for automobile theft hot spots in Austin, Texas.

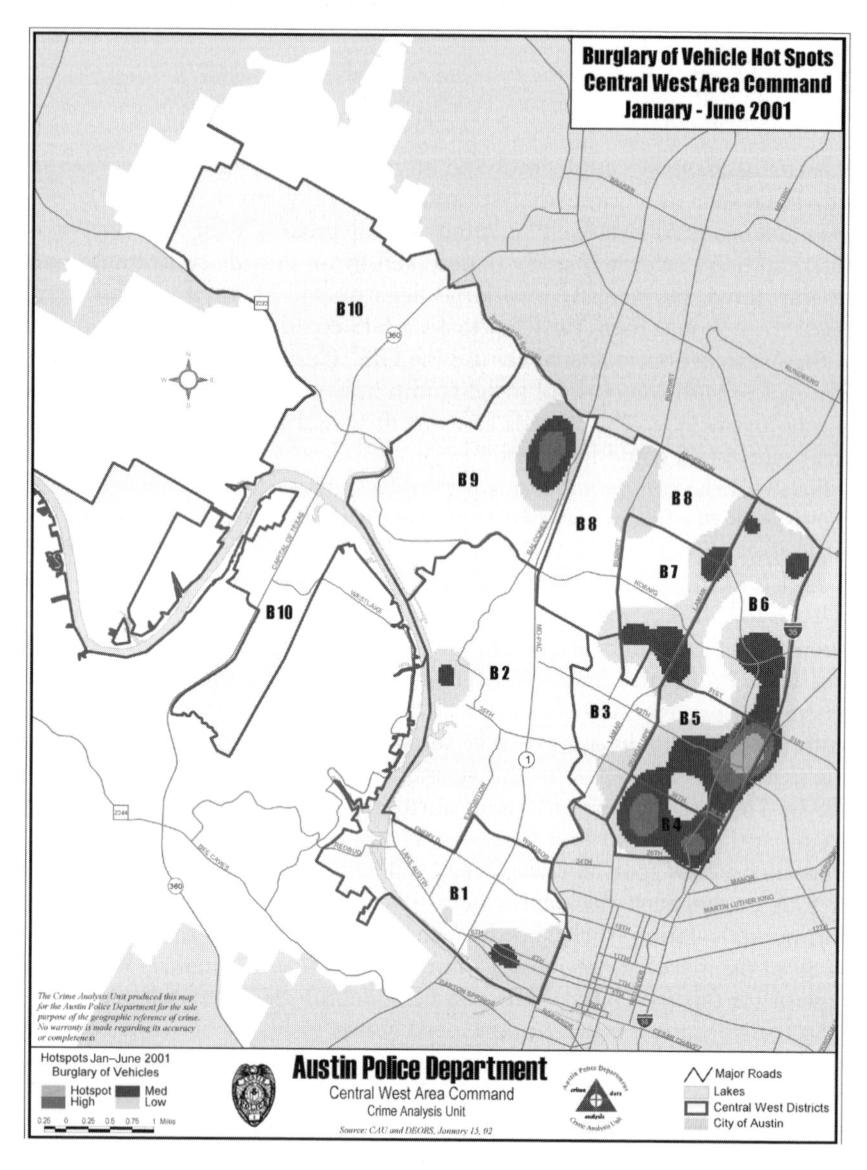

Figure 2.2 Crime analysis of burglary from vehicle hot spots in the City of Austin, Texas using the ArcView GIS from ESRI. Here a crime analyst for the Austin, Texas Police Department uses ArcView to display and analyze burglary from vehicle hot spots for the central west area command for the period January to June 2001. *Courtesy of Sue Barton, Assistant Director, Austin, Texas Police Department.*

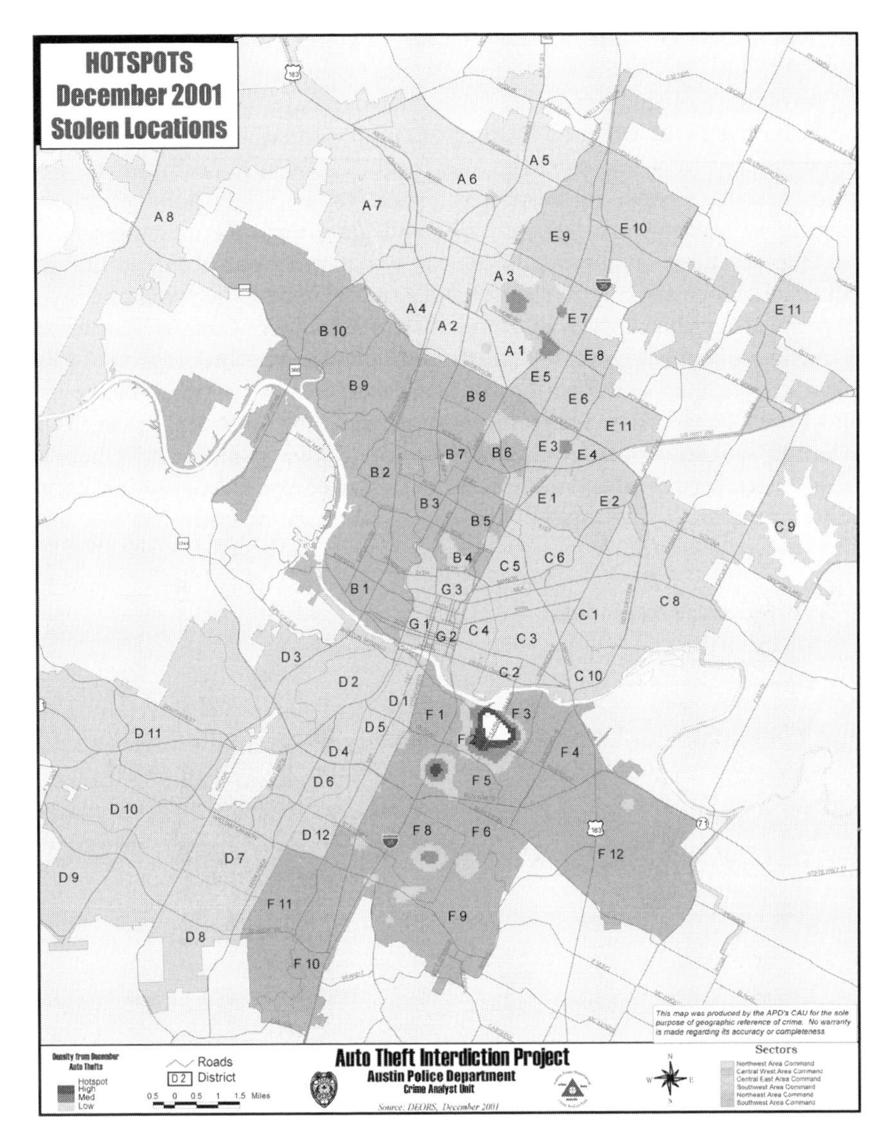

Figure 2.3 Analysis of vehicle theft hot spots for the City of Austin, Texas. Vehicle thefts are concentrated around several hot spots. One is in the vicinity of the "6th Street" area that is known for its numerous nightspots and bars that are patronized by young people, particularly on the weekends. On-street parking in the area is common in the late evening and affords thieves easy opportunities. Another hot spot is close proximity to a motel noted for harboring transients, etc. Awareness of features in the community conducive to crime can help guide analysis of crime data and response efforts. *Courtesy of Sue Barton, Assistant Director, Austin, Texas Police Department.*

Thirdly, specific enforcement *task forces* within a department, for example, narcotics, vice or traffic might need ongoing data on the pattern of incidents and location of features in the community relevant to their responsibilities. Thus, a monthly "map" of arrests for solicitation with the location of sexually oriented businesses and "hot bed" motels highlighted, might be used by a vice squad in San Diego, California. Locations of narcotics arrests, known or suspected "rock houses," locations of public (Section 8) housing projects, schools, and gang territory or gang-related incident locations might be useful for a narcotics enforcement unit in Salinas, California. Gang territories and gang-related arsons would be helpful to investigation of over-all criminality in Winnipeg, Canada. Locations of traffic accidents (Figure 2.4), traffic enforcement activities (check points, stops, arrests, etc.) and locations of licensed dispensers of alcoholic beverages might be useful for a traffic enforcement unit in Tyler, Texas, as it would be in Antwerp, Belgium or Reykjavik Iceland or almost anywhere in the Industrialized world. (Figure 2.5). In the latter example, maps or statistical summaries in the form of bar-charts of traffic-related problems and

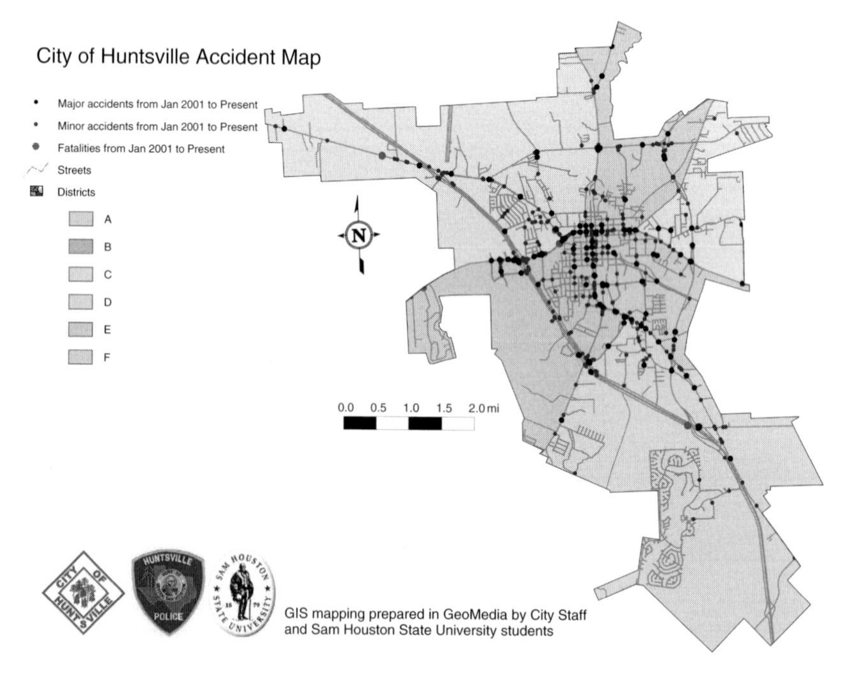

Figure 2.4 Map of automobile accident locations within the City of Huntsville, Texas for the year 2001. This map was generated using the Intergraph Geomedia GIS. It shows locations of vehicular fatalities, and major and minor automobile accidents. Students at Sam Houston State University worked with the Huntsville Police Department and the GIS and MIS staffs of the City on this ongoing effort to identify and reduce traffic accidents. *Map courtesy of William Villines, GIS Manager, City of Huntsville, Texas.*

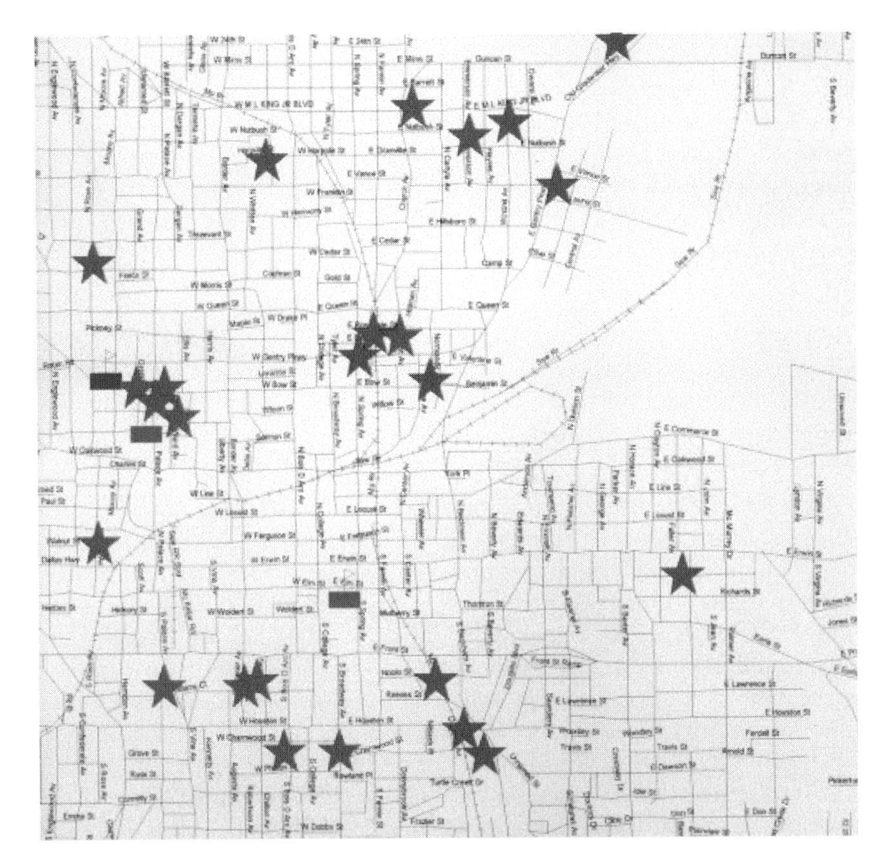

Figure 2.5 Portion of a MapInfo generated map showing street centerlines and alcohol-related automobile accident locations for August 1998 in the City of Tyler, Texas. Smith County in which Tyler is the county seat and largest city is "dry" (prohibits sale of beer, wine, and spirits except in limited circumstances). The flags are "private clubs" (first drink is $10 and includes a membership card) that serve liquor and restaurants that serve beer and wine. *Courtesy Ken Findley, Assistant Chief of Police Tyler, Texas Police Department.*

enforcement actions highlighting the time of day and day of the week would be particularly important.

Lastly, at the *patrol level*, officers in a given precinct (district) or assigned to a specific beat might need hard-copy maps and summary reports of any of several classes of major crimes and possibly less serious crimes and infractions occurring in the pervious week, day, shift, etc. for their own beat or precinct and possibly adjacent spatial areas (Rich 1996). All of these needs would be identified and assessed, independent of a specific GIS-based solution in the *needs assessment* step of the implementation process.

Assignment of responsibilities for system development

At this point in the implementation process, a particularly important decision must be made. Where will responsibility for the GIS reside in the organization? This can be a very problematic issue. In the case of a single crime analyst wishing to periodically use the GIS as an additional tool in his or her crime analysis arsenal, the issue is minor, although development of an adequate base-map and ensuring accurate and timely geo-coding of incident locations is still a challenge. However, in many organizations there may exist an information systems manager or database administrator or crime records manager or other information technology (IT) professional that will feel duty-bound to get involved in the GIS implementation process. Unfortunately, very few such individuals will do an adequate job of developing and implementing a GIS, without additional training and education. This is partially because GIS presents problems, which most information technology professionals have had no experience in solving. All too frequently, local governments including police organizations do not offer competitive salaries sufficient to retain or attract highly qualified information technology professionals and instead have relegated IT jobs to individuals that can barely maintain existing obsolete "legacy" systems. In many instances, existing IT personnel in the organization do not have an adequate background to understand fundamental GIS-related issues such as geo-coding, selection of appropriate base-maps, conversion between, and selection of coordinate systems, projections, and datums and a host of other geo-spatial issues. In some instances, in order to avoid assuming an additional unwanted burden, they may insist that a GIS is unnecessary since a word processor can print out a scanned map along with charts imported from a spreadsheet or that the new computer-aided dispatching (CAD) system is really a GIS since it, in effect, geo-codes 911 calls on a base-map of some sort, which can then be viewed or printed out. For an example of a computer-aided dispatching system that is built on a GIS foundation and has advanced spatial analysis and mapping capabilities, see Figure 2.6. Interestingly, in contrast to many IT people, crime analysts, sworn officers, and other end-users of GIS in law enforcement rarely are antagonistic to adoption of a GIS, but may have unrealistic expectations of the ease of implementation and the need for allocation of resources such as those involved in geo-coding incidents on an ongoing basis.

A better alternative than relying solely on "in-house" expertise within the police department (in those cities or counties that have a GIS function in a public works or comprehensive planning department) is to use the expertise of those other departments to get the GIS up and running in the police department, then gradually have the police department assume increasing responsibility for the system. Examples of police departments successfully using this approach include the metropolitan police of Las Vegas, Nevada, and the city police of Ontario, California and the Pinellas County, Florida Sheriffs Department.

Figure 2.6 Central Dispatching Center for the National Police of Iceland in Reykjavik. This computer-aided dispatching system uses ESRI software customized by HNIT Consulting Engineers in Reykjavik to manage, display, and accurately geo-code calls and store associated attribute data for calls for service made to the national 112 emergency number. These are stored in a format available for simple incorporation in a GIS program, in contrast to the output of most CAD systems in use by police agencies and emergency communication districts in the world today. *Photo made available for publication in this book courtesy of Jon Svavarsson, MOTIV-FOTO, Reykjavik, Iceland. © MOTIV-FOTO, Jon S.*

As the scope of GIS use within a police department increases it may be possible to develop a dedicated *GIS analyst* position. This professional may possibly be assisted by one or more *GIS technicians* and eventually, in the largest organizations, a whole GIS department may develop with a manager, several analysts, and numerous technicians. In any case, the options of assigning responsibility for the GIS to a crime analyst or an interested sworn officer and/or utilization of existing GIS capabilities within the government entity and/or hiring or internally developing a *GIS analyst* are likely to result in a more satisfactory outcome than relying solely on an existing database manager, crime records administrator, or similar functionary. Exceptions do exist however, and a particularly gifted and interested information technology professional may be the ideal choice to guide development of the system, with the understood caveat that this individual will require both training and education in GIS-specific issues.

Functional requirements study

The third step in the implementation process would be a study of the manner in which GIS would meet the departmental needs for geo-spatial analysis and data management, which were identified in the needs assessment. In order to undertake such a study, some input from GIS vendors concerning their software, and its functionality, would be desirable. For example, this *functional requirement study* step would help the decision maker identify whether the GIS would need to be interfaced with relational database management software or an existing crime records management system or if it would function on a stand-alone basis. This study would help identify what analysis capabilities such as exception reporting, buffer-zone generation, spatial statistical functionality, automated geocoding network analysis and other higher-level analysis functions would be required. This step would help determine if the requirements of the department could be met with a single copy of a "desktop mapping" program, multiple copies of a "high-end" GIS capable of creating and maintaining GIS data from scratch, or by adoption of a customized GIS program designed to perform higher-level analysis specific to crime mapping. In the first case, a single copy of software like MapInfo from MapInfo Corporation Inc. would be appropriate (Figure 2.5). In a differing situation, use of a single additional copy of the Geomedia GIS from Intergraph Corporation in Huntsville, Alabama (Figure 2.4) might best allow an existing city engineering department's GIS to be utilized in the police station by the municipal police department. In another city, a copy of ARCGIS plus multiple copies of ArcView from the Environmental Systems Research Institute (ESRI) in Redlands, California, might meet the law enforcement organizations needs (Figures 2.2 and 2.3). Some organizations have chosen to utilize the MapGuide GIS product from AutoDesk Corporation, which can integrate GIS capabilities with the popular AutoCAD program (Figure 2.7). In yet another situation, a program like *CRIMEVIEW* from The Omega Group of San Diego, California might be employed. This software developed by a third party developer based on ESRI products and customized to a particular department's crime mapping and analysis needs might be most appropriate (Figures 2.8–2.12). Last but not least, a crime analyst might choose a software product such as the Rigel geographic profiling software from Environmental Criminology Research, Inc. of Vancouver, British Colombia, Canada to address a specific problem such as serial crimes to which it is particularly applicable (Figures 2.13–2.15).

Selection and configuration of hardware and peripherals

Once a set of functional requirements and a GIS program whose capabilities satisfy those requirements has been identified, a decision must be made as to what hardware and peripherals are needed to support the input,

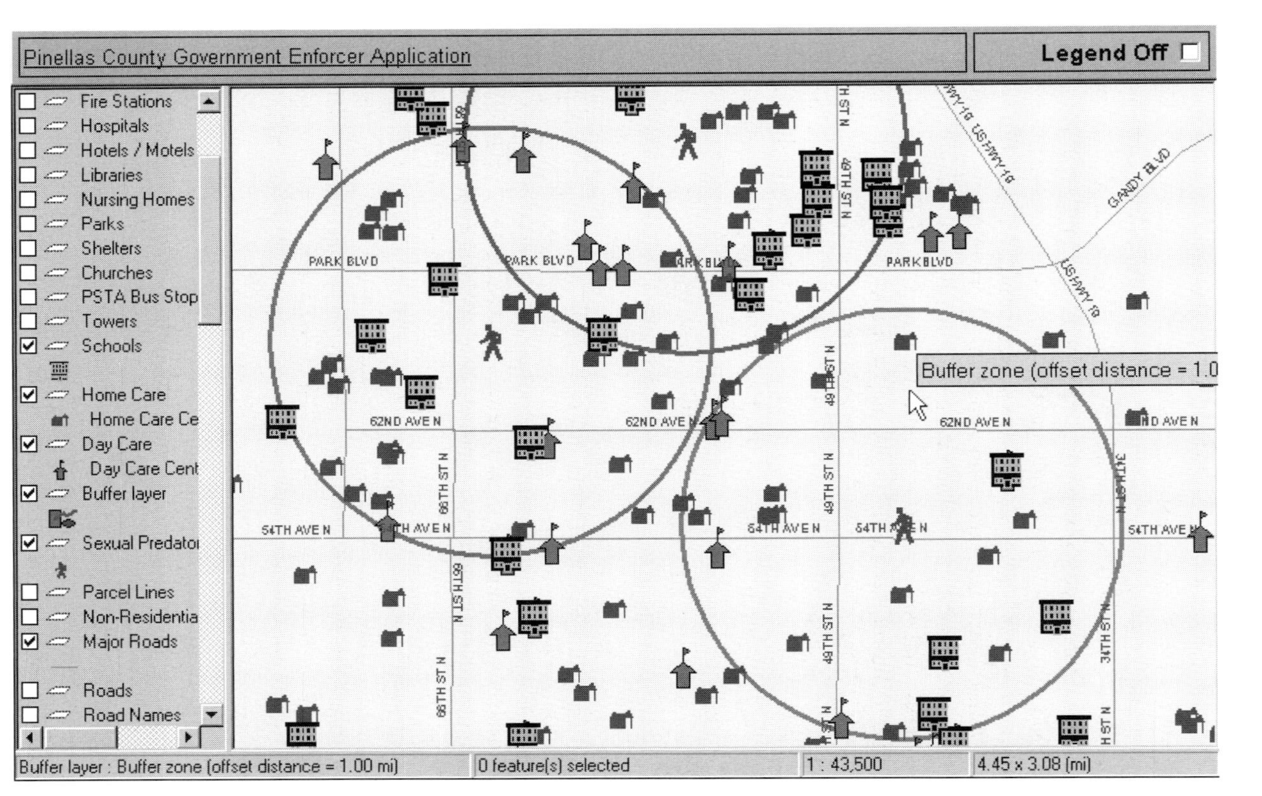

Figure 2.7 Use of buffer-zone analysis in law enforcement is illustrated by this map created by the Pinellas County, Florida Justice Coordination Department, using the GIS software from AutoDesk, Inc. This map illustrates all the schools, day care centers, public parks, and other features of concern within a 1,000-foot proximity of the residence location of a registered sex offender. *Courtesy of Tim Burns, Pinellas, County, Florida Justice Coordination Department, Clearwater, Florida.*

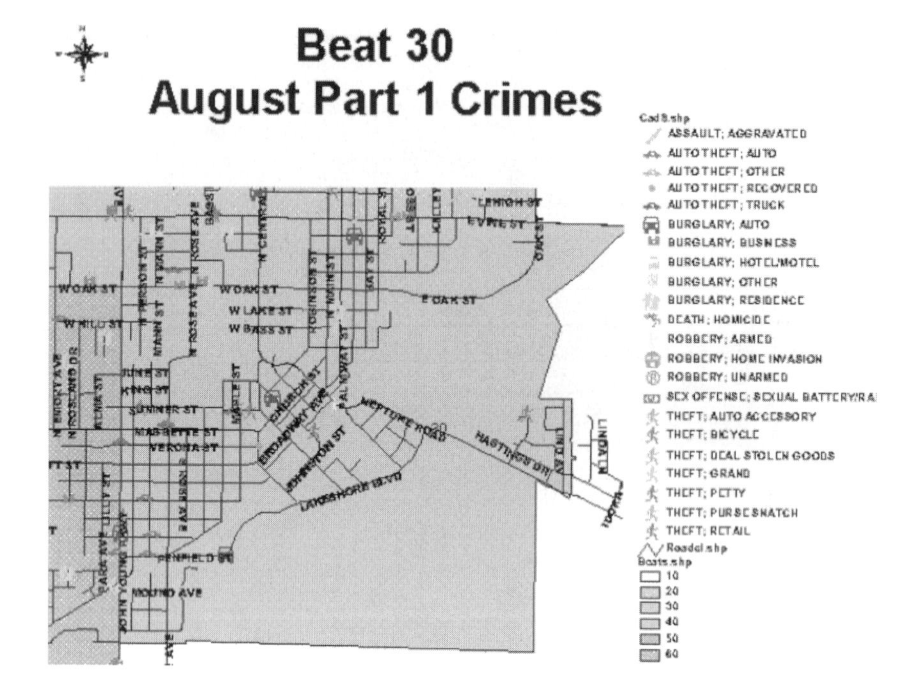

Figure 2.8 Use of the CRIMEVIEW® by the Redlands California Police Department is portrayed. CRIMEVIEW is an extension to ArcView developed by The Omega Group of San Diego, California. Here, rather than use a single geometric symbol or multiple shapes or even graduated symbols for a map showing all major (part 1) crimes for August 1999 for Redlands California, Omega uses a customized crime mapping symbol set (available over the Internet from the ESRI web site: http://www.esri.com/...) Symbols using various colors and connotative shapes have been developed and can be automatically invoked or accessed for manual placement from a marker symbol set. *Note Figures 2.8–2.12 are all courtesy of Milan Mueller, President, The Omega Group Incorporated in San Diego, California.*

computation, and product generation (plotting) aspects of the system. This is an increasingly simple matter, as most reasonably current personal computers (PCs) are fully adequate to support most GIS programs performing typical law enforcement applications, accessing typically sized data sets. However, a larger than standard monitor is desirable (21–24 inches), as is an enhanced graphics card (if an earlier generation of PC is in use). For departments managing large data sets, hardware such as large capacity hard drives and CD writers and possibly servers with enhanced storage capabilities are usually desirable. Most police departments find that a GIS running in a client-server environment under an operating system such as Microsoft

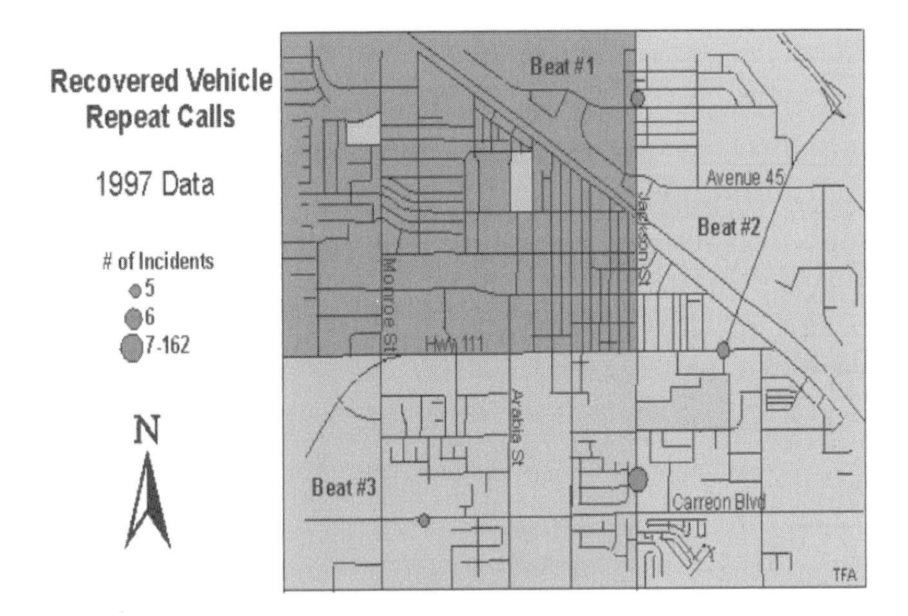

Figure 2.9 This map, generated using the CRIMEVIEW® software, shows locations of recovery of stolen vehicles for Redlands, California. This is portrayed using a graduated size dot. The larger the dot is, the greater the number of incidents or calls for service at a given location. This approach is particularly useful for locations where incidents are concentrated. Consider disposal of "stripped" cars near a "chop-shop," or a particularly troublesome intersection for automobile accidents or noise, disturbance or public intoxication calls, at a particularly rowdy bar, apartment house, residence, or fraternity. Once the dot begins to take over the map, concerted action may need to be taken to reduce further repeat calls or identify the proximal offender.

Windows NT or 2000 etc. is most efficient, although many departments have a GIS running on a stand-alone computer. Some departments that have not yet implemented a GIS may find that the existing crime records management system is on a mainframe or mini-computer with a proprietary operating system and programs written in obsolete languages such as COBOL. The difficulty in integrating this "legacy" system with a GIS is not so much a limitation of the GIS, as it is yet another indication that this data management approach is obsolete. In the more sophisticated and early adopting police departments, the GIS may be using a single (or networked) UNIX workstation. In those departments with more limited resources it may be on a stand-alone PC, sometimes only running Microsoft Windows-based programs. In either case, a successful implementation of the technology can be achieved.

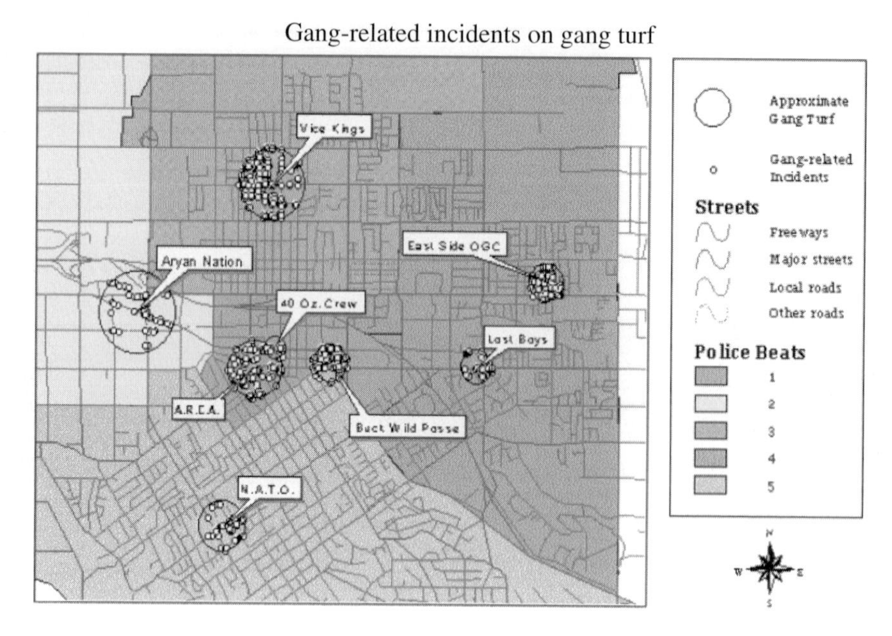

Figure 2.10 This is an example of specialized analysis feasible using GIS. Gang-related incidents have been mapped and hot spots generated and encapsulated by a buffer-zone generation function of CRIMEVIEW®/ ArcView Spatial Analyst. Note the creative names such as Buck Wild Posse and 40 Oz. Crew of actual gangs active in Redlands, California. A map showing "gang turf" might not be one that some members of the community (such as the chamber of commerce) would like to see widely disseminated, however useful it may be for crime analysis within the Police Department itself.

Input devices such as optical scanners are very useful, although large format ones may prove prohibitively expensive. The value of digitizer tablets is limited to those departments that intend to develop their own base-maps through traditional digitizing. A better alternative is to have a *service bureau* or digital conversion facility use a large format scanner and provide either the vectorized output or at least a raster image that can be "heads-up digitized." Ironically, prison inmates in Texas, Oklahoma, Florida, and California are being used as "*digislaves*" in such heads-up digitizing *production shops*, although thus far they have been principally tasked to convert state highway maps or city and county parcel maps from paper or CAD formats into a topologically structured vector GIS. In an odd twist of fate, the updated street centerline GIS data being created by inmates may help to incarcerate their brethren on the outside or help recapture recidivists among their ranks.

Essential output devices include a dedicated color printer. High-resolution ink jet printers are now very affordable, although for large volume

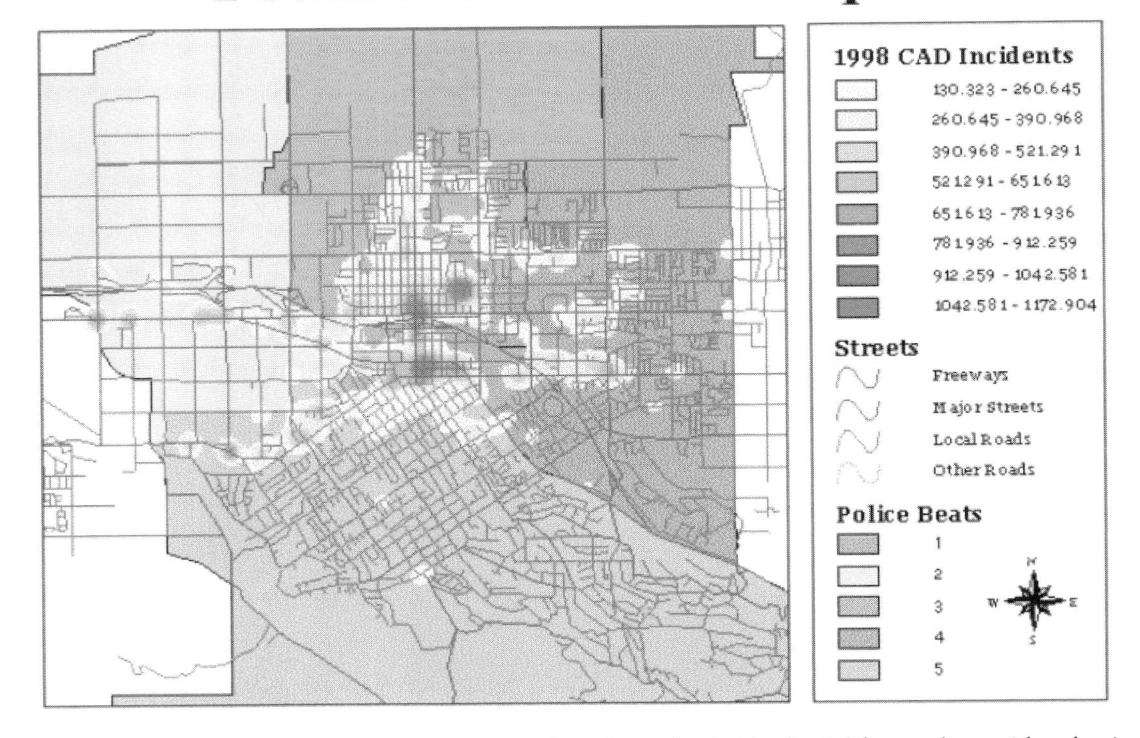

Figure 2.11 Here, we see the street centerlines and police district boundaries for Redlands, California along with a density surface generated by CRIMEVIEW®/ArcView Spatial Analyst. This surface is much like an elevation contour map and shows the greater concentration of calls for police services made to the CAD (computer-aided dispatching emergency 911) system. It provides an overall view of police department calls for service rather than locations of specific incidents. When there are large numbers of incidents to be displayed such an approach rather than use of points or even graduated symbols may be more informative.

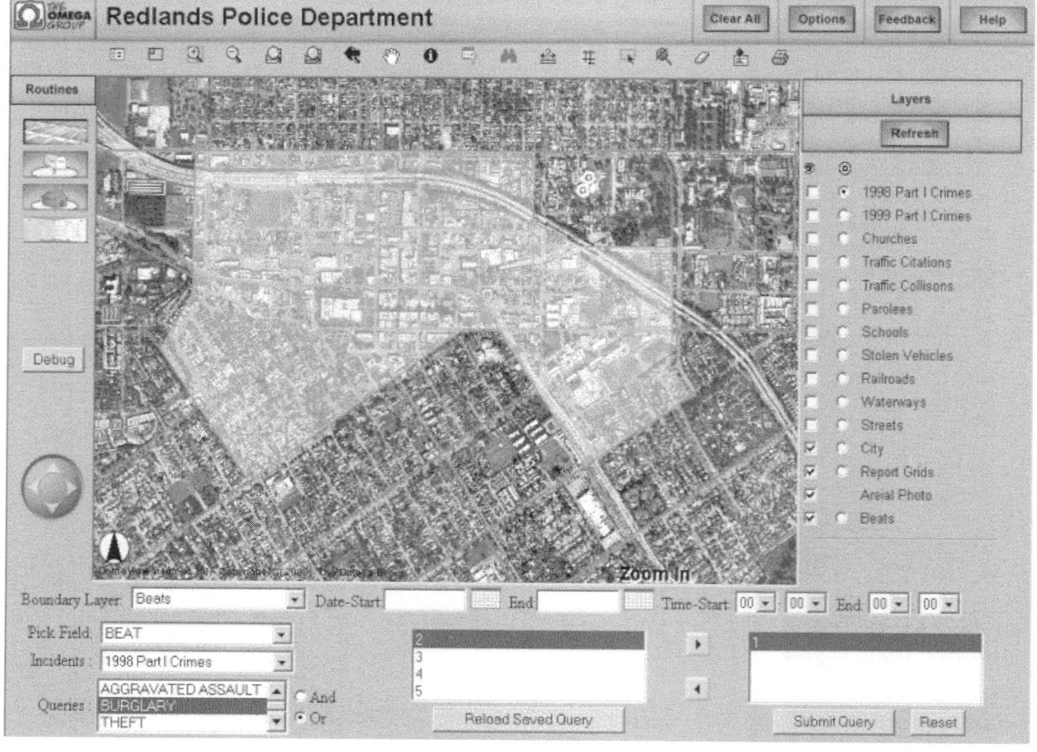

Figure 2.12 Use of digital ortho-photography as a base-map rather than typical street centerlines and jurisdictional boundaries is shown on this image. The graphical user interface design for CRIMEVIEW® is also displayed. Here, a district in Redlands, California is delineated with a transparent fill on top of the raster-based digital aerial ortho-photography. Incident locations are highlighted by point symbols. Such digital aerial photography, if available, can help to fill in the gaps between street centerlines and is an excellent vehicle for public presentations. For experienced users however, the street network usually provides an adequate context for decision-making.

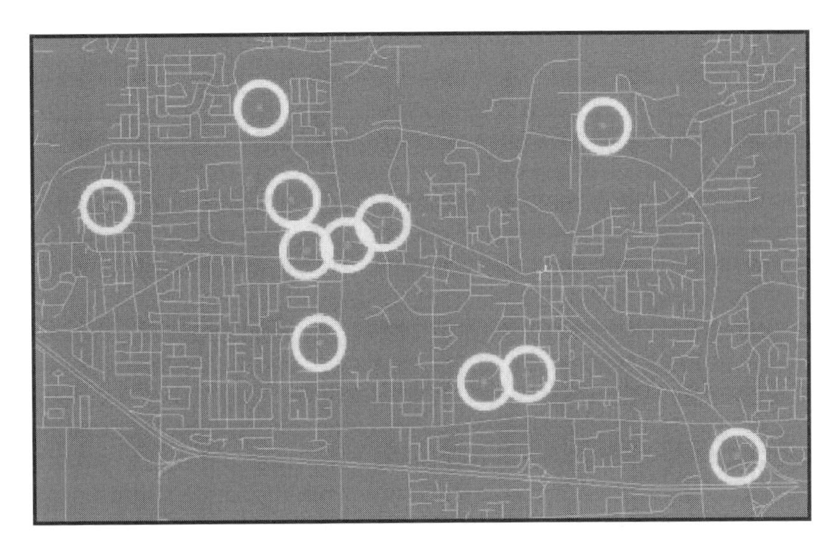

Figure 2.13 A base-map of the streets in Abbotsford, British Columbia, Canada along with locations and buffer-zones associated with attacks and other known activities of a serial murder suspect. Generated by Rigel Software from ECRI Inc. Courtesy of Ian Laverty, President Environmental Criminology Research Inc Vancouver B.C. Canada.

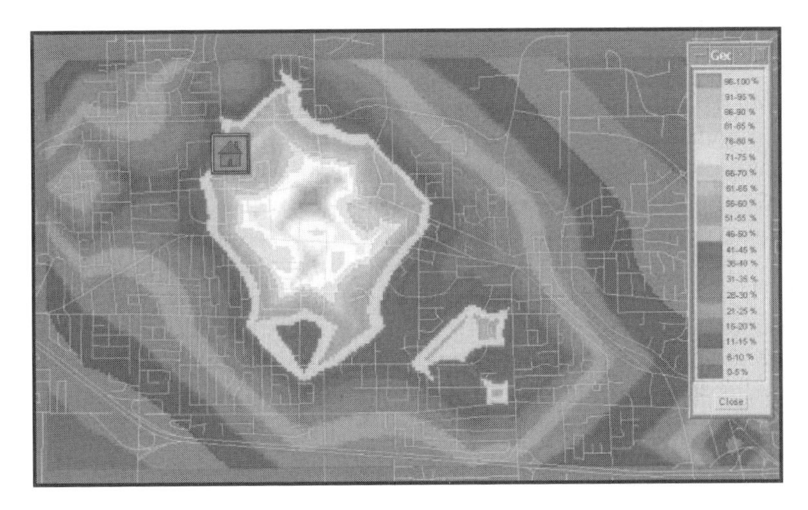

Figure 2.14 A two-dimensional probability estimate over a street center-line base generated by the Rigel software of the *activity space* of the serial murder suspect. It helps to identify areas where the suspect is likely to strike again if he adheres to his established pattern. The image also shows the location of the residence (small house symbol) of the individual ultimately determined to be the serial killer. Note that there is a close proximity between the killer's residence and the area where it was estimated the highest probability of additional attacks was likely to occur. Predators may not reside at the center of such probability analysis "hot spots," but they frequently live or work in close proximity to where they commit some aspect of their crimes.

Figure 2.15 A "3D" surface representation of this same probability analysis with street centerlines draped across it derived from the probability analysis shown in Figure 2.14. Such a surface can help to better highlight areas of highest probability, compared to traditional two-dimensional *isoline* (contour) type maps.

printing (such as generating daily crime maps for all patrol officers), a color laser printer might prove cost-effective. A very useful peripheral is a large format plotter such as a Hewlett-Packard, ENCAD, or similar *E size* color inkjet plotter. These can be used to generate large format maps, banners, and other materials for public presentations and display in the offices and public spaces of the department. These output devices can really help to raise the visibility of the department's crime mapping activities. They are an investment that can help justify the man-power commitment and sheer drudgery involved in many facets of the effort to develop a GIS to a sometimes skeptical public and/or to civilian or internal oversight groups.

Some other useful peripherals include video projectors and laptop computers for making internal and external presentations. In some departments (such as the New York City Police Department) video projector based presentations are used during monthly command level briefings to display the results of GIS-based analysis. In New York City this has become a key part of the COMSTAT (comparative statistics) decision-making process.

For many departments, GIS has been integrated with global positioning systems (GPS) and/or automatic vehicle locator (AVL) systems. Examples include the Corpus Christi, Texas, and Ontario, California, Police Departments, the Pinellas County Sheriff in Florida and the Illinois State Police (Figure 2.16). In some departments such as that of Ontario, California, the GIS has also been implemented on mobile data terminals using Mapobjects, a minimalist GIS from ESRI. These technologies can also

Figure 2.16 An example of how information for a GIS can be gathered in the field. This "ruggedized" laptop is part of a system that utilized a GPS unit built into Illinois State Police patrol cars to rapidly determine the location of incidents on State highways, Interstates, and other roads patrolled by the ISP. These incidents are entered into a template stored on the laptop along with GPS-determined locations. These are then incorporated into a GIS at the State Police Headquarters in Springfield for analysis in issues such as drug interdiction, serious accidents, drunk driving stops, and other enforcement-related issues. Eventually, the GIS itself may be available on the laptop with real-time updated information. *Photograph courtesy of Sam Nolen, Director of the Illinois State Police, Springfield, Illinois.*

be rightly regarded as useful input devices for a GIS system, or the GIS can be seen as a display and analysis device adjunct to the mobile systems. Whichever concept is embraced, in such a combined approach, GIS must be regarded as an integral component (Sorensen 1997). GIS and GPS and other sophisticated communications and electronic technologies can also be taken into field situations in specially configured vehicles designed for collection and analysis of geo-spatial data such as the Crime Analysis Response Team (CART) mobile command center of the San Bernardino County Sheriff's Department in California (Figures 2.17 and 2.18).

Base-Map creation

Once a GIS program has been selected and installed on an adequate hardware platform and operating system, running in an appropriately networked setting and connected to relevant output devices and peripherals, the task of creating or obtaining an adequate base-map must be undertaken. For most law enforcement agencies, the minimum requirements for a GIS base-map are *street centerlines* for the city or locality in question and the

Figure 2.17 This is an exterior view of the San Bernardino County Sheriff's Department's Crime Analysis Response Team mobile crime analysis unit. It has capabilities for wireless Internet and cellular communications, crime mapping, and analysis using ArcView and GPS. *Photograph courtesy of Michelle Smith, San Bernardino County California Sheriff's Department.*

Figure 2.18 Interior view of the San Bernardino County California Sheriff's Department's mobile Crime Analysis Response Team unit showing computer equipment that can be deployed to any point in the county. San Bernardino County happens to be the largest county in America covering over 19,000 square miles in surface area, mostly sparsely inhabited desert, so a mobile capability is important. *Photograph courtesy of Michelle Smith, San Bernardino County California Sheriff's Department.*

boundaries of relevant jurisdictional areas. The streets (stored in the GIS as a series of line segments sometimes referred to as *arcs*) may possibly be differentiated by color and line type as to kind (local, arterial, divided highway, etc). In the newest generation of object oriented GIS programs such as ESRI's ARCGIS 8.X products behaviors can also be associated with street segments or whole streets stored as a single feature. The jurisdictional boundaries stored as polygons, would include, at a minimum, city limits, county or state lines (if present), and precinct (district) and usually beat boundaries. In some cases, the boundaries of city council or county commissioner districts, justice of the peace jurisdictions and US census subdivisions such as tracts might be useful, particularly for re-aggregating crime incidents to present locations and rates to public officials or to explore the demographic factors related to crime occurrence (US Census Bureau 1997).

Other features relevant for the GIS would be the police department's own location (possibly a point or alternatively a polygon), substation or precinct locations (if present), city hall, fire stations, hospitals, schools, and public parks (both frequently a foci for crime incidents). Other features that have been added to some law enforcement base-maps include jails, community supervision facilities, mental health and drug treatment facilities, liquor stores, bars, sexually-oriented businesses, motels and single resident occupancy hotels, pawn shops, convenience stores, traffic signals and even street lights and other specific features that may relate to the environmental factors contributing to the incidence of crime in a community.

Obtaining the location of such features can be a challenge, but GPS offers a low-cost method to determine the absolute locations of such features relatively easily. Geo-coding business licenses has also been tried in a few large cities, but does not work very well because owner addresses may not correspond to the physical location of the business establishment. If a particular feature in the community has a law enforcement issue, it may warrant an effort to selectively include that feature in the GIS. For example, in a county in north Texas, church arson and vandalism of graveyards was a high priority issue. A day or two of fieldwork by an officer equipped with a GPS unit would suffice to map the location and boundaries (as an area feature) of all churches and associated cemeteries in the county. Similar efforts to map parks and bicycle paths have also been undertaken. Some local and regional government agencies also have data on land-use, zoning, parcel maps with building foot-prints and even interior layout for commercial and industrial buildings (usually in a computer-aided design format and primarily used by fire departments). Also many local and regional planning agencies have digital aerial ortho-photography that can be used as a "backdrop" to help "fill in the blanks" in the space between the street centerlines in the GIS. An example of such an application is used by the Detroit, Michigan Police Department, which finds it particularly useful to help identify burned-out buildings, vacant lots and alleyways that are frequently linked to crime but are not particularly easy to identify on street centerline type GIS base-maps.

Geo-referencing issues

In order to use this GIS base-map to provide a context for the display of crime incident locations, the base-map must have a coordinate system, projection, and datum. In some sense, selection of "*the ideal*" set of geo-referencing characteristics is less important than consistency. That is, if state plane coordinates are being used (for instance by a city's public works department) sticking with that coordinate system and its associated projection and datum in all geo-spatial development activities in city government is what is most important. Data arriving in some other set of coordinates will have to be cartographically transformed before it can be combined with the existing data in the same GIS. A department with responsibilities over a wide area such as a State Police Department may not choose to use a state plane coordinate system because there may be multiple zones within a state, thus a system like Universal Transverse Mercator might be preferable. Some state agencies such as the Texas Department of Transportation have even taken the step of developing their own coordinate system, a single zone covering the entire state using northings and eastings in feet. Since this system is not compatible with any other system in common use, it has many drawbacks. In very few cases would a system like latitude and longitude be appropriate, and then only for law enforcement agencies such as the US Coast Guard dealing with marine navigation issues, and comfortable with the lack of a Cartesian frame of reference (Cambpell 1998).

Selection of a projection may be predetermined by selection of the coordinate system, such as the one used for the prevailing state plane system. If it is not so specified, then once again the specific choice is less relevant than consistency in sticking to that choice and conversion of input data into that projection. Worth noting is that most computer-aided dispatching (CAD) and computer-aided drafting (CAD again) data will lack a projection, since the generalized base-maps used are planimetric and also that all projections create distortion. If an analyst wishes to perform any calculations in which the area of features is employed, then selection of an equal area projection is essential to avoid errors in estimates such as those for population density or density of crime incidents in a given area. However, these sorts of estimates can be misleading. For example, the regional economic development authority of a leading American city generated a GIS-based analysis that indicated that the impoverished Southeast side had a "lower than average" crime density on a per square mile basis. The fact that much of this area was uninhabited flood plain rendered this "crime density" analysis highly misleading, no matter how much it may have reassured potential investors in re-development efforts in this area. Larger spatial areas cause the consequences of not using a projection to be more pronounced, thus for a small city, using a set of CAD data without a projection will produce few problems, while for a national law enforcement agency, use of an appropriate projection is essential for accurate spatial analysis.

As with coordinate systems, selection of a uniform datum is more important than whether NAD 83 or NAD 27 is selected. If data that was collected earlier than the middle 1980s is used, it will require transformation prior to incorporation into a system utilizing the newer North American Datum (outside North America other datums are available).

Geo-coding issues

Most police departments do not gather exact geographic coordinates for crime incidents or for calls for service (911 calls in most of the US). Rather, a street address will be recorded by a responding officer or taken by a dispatcher or automatically generated by a computer-aided dispatching system's database linking phone numbers and addresses. In any case, the street address is not an exactly accurate coordinate whether it is recorded by a responding officer or interpolated to a point on a base-map by a CAD system. Such approximate street addresses for incidents are the bane of GIS development efforts in law enforcement agencies. Frequently, the incident reports taken in the field are found to contain spelling errors, "guesstimates" as to correct address and mysterious abbreviations or notations like "old-town 711." The officer making the report may know where the "old-town 711" convenience store is located, but the geo-coding function in most GIS systems is baffled by this type of input data. Incidents not occurring at a readily identifiable street address are a particular problem in rural areas. Thus, a body recovered from "next to Boggy Creek on farmer Shultz's south 40" cannot be geo-coded. A GIS geo-coding algorithm would also typically reject streets given colloquial names or having multiple names along various sections. To deal with colloquialisms in incident reports, a data dictionary that "aliases" various renderings of place names can be used (as in the City of Salinas) where all convenience store names and corresponding locations are so aliased.

Computer-aided dispatching systems cannot cope with many newer addresses or accurately locate calls coming from cell phones (although triangulation and/or GPS of cell phone caller locations is a rapidly approaching reality). Also a very high proportion of 911 calls never result in generation of an incident report. Multiple calls will be received about the same incident from many addresses ("Shots fired nearby") and calls may be misdirected ("Can you give me the correct time…") or simply nuisance calls ("My neighbors are actually Martians…"). Perhaps most importantly, numerous incidents are discovered by officers on patrol. Making the officer put a call into the 911 system from a phone near the scene seems an absurdly indirect way of geo-coding crime incidents (Canter 1998).

Confusing and multiple street names are commonplace. To cite an example from Huntsville, Texas: "Possum Walk Road" exists only in the minds of many long-term residents, officially it is "Farm to Market Road # 1374" and it changes to "Old Montgomery Road" for part of its length. A geo-coding look-up table and associated interpolation algorithm built into a GIS would fail to recognize any incidents reported on "Possum Walk Road" and if the address for the incident fell into the range where FM 1374 is officially "Old Montgomery Road," then the geo-coding would either fail, or conceivably it would locate the point representing the incident location in the wrong stretch of this road. Other common examples of problems that affect geo-coding are multiple streets with the same name (there are five separate streets named "Lake Road" in Walker County, Texas, several in the same city) or streets with names that are confounded with abbreviations stored in a data dictionary in the system. Thus, for an incident at 2301 Ave. S, the "S" might be assumed to stand for "South" and the system would expect a street name to follow. Even "A." or "Av." or "Ave." could cause a problem, if the system expected to see it spelled out as "Avenue" or if the period indicating use of an abbreviation were omitted. Similarly, if a cardinal direction is not specified, the incident may not geo-code properly, since there may be two "7,300 Lakeshore Drives" in Chicago that are miles apart (one at 7,300 South and one at 7,300 North).

Geocoding "wizzards" now exist with fuzzy logic that enables them to accept minor spelling errors and common variations such as Av. or Ave for avenue. A lack of care in obtaining, recording, or in key-punching the corresponding address for an incident is frequently demonstrated by police officers or data-entry staff, particularly those unfamiliar with the uses to which the collected data will be put and the likelihood that an erroneous address will render the collected information of little value for use in a GIS. The requirement that a sergeant or shift supervisor "sign off" on incident reports prior to entry into a database or development of an automatic error checking capability with rapid follow-up of reports rejected as incomplete can help improve geo-coding efficiency.

Other complications may result because of incidents that occur at more than one location. Take for example, the case of the abduction of a policewoman from the Greenspoint (locally called the "Gunspoint") Mall parking lot in Houston, Texas. The woman was abducted at one location; murdered at another; her car was recovered at a third location; and her body at a fourth location; and the perpetrators had several (possibly false) home addresses. Which address should be tied to the incident; should this single crime generate multiple point symbols representing homicide? An object-oriented GIS could treat all locations as a single object and multiple records could be linked together in a relational database management system, but in reality most police departments will settle on a single point with notations in the incident report that other geographic locations are tied to that one incident.

Another difficulty is presented by the lag between the incident report, the resulting arrest (if any) and the ultimate legal disposition of the case. If incidents are geo-coded promptly, then the crime reported will often be more serious than the ultimate charge. For example, in a *Santa Rosa*, Northern California, case the initial charge in the incident report taken at the scene was "attempted murder" and "grand theft auto." Only later was a plea bargain struck to reduce the charges to "reckless driving" (a man had run over his girlfriend as he fled in her car at a high rate of speed after she had tried to kill him and then smashed the windows and doors on the car with a tire iron as he tried to flee). While it is valuable to track not only where crimes occur, but which crimes are cleared, waiting months for the ultimate disposition of cases with the need to keep in contact with the district attorney's office etc., is more effort than many departments want to take to maintain a truer picture of the incidence of crime.

Geo-coding is an ongoing headache for most entities building a GIS using street address data. This includes market researchers geo-coding customer information, which accounts for their frequent use of zip codes or phone prefixes rather than street addresses for geo-coding. Unfortunately, these options cause a generalization of the data that is unacceptable for law enforcement purposes. Those cities that are systematically laid out in grid-iron street patterns with a uniform population density, where addresses can be easily interpolated, are in the best shape for geo-coding. Those localities which, for emergency dispatching purposes, have insured that all street addresses are unique and that all residences have an address that correlates with a specific geographic location (rather than a P.O. Box or rural delivery route number) are in reasonable shape. Suburban areas with curvilinear streets and non-uniform population density pose a challenge, while areas without determinable addresses pose an even larger hurdle. Unless GPS is to be used to collect data on incident locations, assignment of a unique address to every residence and business is necessary for both GIS development and efficient provision of emergency services. Finally, areas undergoing rapid and unregulated growth insure that available GIS base-maps such as the US Census Bureau's TIGER (topologically integrated geographic encoding and referencing) files and even the enhanced versions of this GIS information available from GDT of Hanover, New Hampshire, teleatlasot Menlo Park California or other vendors will miss significant features such as new streets. In particular, areas such as recently built trailer parks on private drives will likely be absent, although they may contain a significant number of crime-prone residents. Thus, stable, well planned metropolitan areas such as Chicago offer the best prospects for geo-coding, while fast growing rural or suburban areas with no comprehensive planning framework and a hodgepodge of addressing systems, rural routes, streets with multiple names, etc. are in the worst shape.

An alternative to using a street-centerline base-map with address ranges such as those created by national governments in the United States (TIGER) and in Australia by the national census authorities is to use a base-map that uniquely identifies parcels of land or building footprints. A system with all building footprints linked to a unique street address offers the advantage that crime incidents at homes, or offices, schools, etc. will be linked to the best possible approximation of the location where the crime actually occurred. Unfortunately, few police agencies (outside Great Britain) have access to such a GIS base-map. However, the creation of cadastral maps by tax assessment authorities can provide police departments with such a resource. An example would be the Pinellas County Florida where the Sheriff's Department geo-codes incidents using a countywide parcel map with building footprints linked to unique addresses. Also some police agencies such as Redlands, California use a municipal GIS that has a unique address and building footprint data.

In all cities the "*hit rate*" (the proportion of correctly geo-coded incidents) is likely to be far less than 100 percent initially. The real test of the successful development of the crime mapping system is whether that rate improves. It will only improve if a data dictionary is created that aliases possible variants of street addresses and if the base-map and associated address-range data is accurate and up-to-date. Furthermore, education of those officers taking incident reports and staff entering the data is crucial. They need to concentrate on correct spelling and endeavor to record the corresponding street number. Also, rules for handling situations like accidents at intersections and incidents in public parks and other locations that lack an easily discernable street address need to be adopted. In the case of public parks, GPS units can be used, the centroid of the polygon defining the park can be employed as a representative "label" point or as in the case of Lakeshore Park in Chicago, a numbered grid index system can be created and printed on maps available to officers to help them locate incidents. The City of Chicago's Police Department has achieved an enviable rate of correct geo-coding on the first pass approaching 99 percent. This has been accomplished by laboriously determining why an incident did not geo-code and making a concerted effort to prevent the same error or class of error from happening again (Block 1995).

However, even in stable gridiron-based cities that have dedicated a considerable effort to base-map development and accurate geo-coding, problems remain. For example, public housing projects such as the Cabrini Green and the Robert Taylor Homes in Chicago may have such a high incidence of crimes that not only are multiple incidents occurring in a single building, or multiple-related incidents occurring in a single apartment, but unrelated incidents are occurring in a given year in as many as ten out of a possible twenty apartments located directly above one another (Block 1997). Thus, the approach of representing each incident by placing a unique symbol at the corresponding incident location on the ground becomes infeasible. The usual solution is to increase the size of the symbol or change its color, a technique used in manual pin maps for many years (Dent 1999). It is worth noting that crime incidents can be differentiated based on the size of the symbol (to represent multiple incidents or highlight more serious ones), by the color of the symbol (such as different colors for differing time periods like months of the year), and by using various geometric shapes. In the case of certain police departments such as that of Redlands, or Salinas, California, unique symbols have been designed that connote various types of crimes. Thus, in the Redlands Police Department's GIS, a robbery is denoted by a running man with a range of colors indicating the specific type of robbery, a home invasion by a sky mask symbol, a stolen or recovered car by an appropriately colored car symbol, etc. (see Figure 2.8). GIS packages have the ability to store libraries of such customized symbols that can be evoked by a user or automatically accessed to insert the appropriate symbol based on contents of the corresponding field in the crime incident record that is being geo-coded.

Education, training and staffing issues

The need to educate patrol officers on the importance of accurate recording of address or other locational information is only one example where training and education play a pivotal role in the successful implementation of GIS. One should differentiate between education and training with respect to GIS. Education is a more general and time-consuming process usually involving universities, community colleges and technical and vocational institutions. To get GIS-related education, a GIS user might enrol in a community college or university course in GIS for a semester. This actually can be a good investment for the department as well as for the individual, although the time commitment can be a major issue in a department already short on staff.

In contrast to education, training is typically software focused and of short duration. All vendors of GIS software provide both on-site and distant training and all also provide Internet-based training as well as numerous CD ROM-based demos and tutorials. As in all types of training, the interest and dedication of the staff being trained is important to a successful outcome. This is particularly true for the *apparently* lower cost Internet and CD ROM-based tutorials. Few individuals really learn effectively from these materials alone. On the other hand, there is no substitute for time spent struggling and learning, in a *non-linear* fashion, the quirks and essential elements of a given GIS software package. The problem is that unless there is someone able to point out shortcuts to efficient use of the software to the neophyte user, the ratio of frustration to progress is likely to be very high. Having an experienced user available to show new users the ropes is ideal.

Concentrated training in the form of vendor- or consultant-provided classes is a good approach. This is particularly true if users have had some opportunity to "play" with the software prior to this training. There are several training centers specializing in GIS training for law enforcement professionals such as the National Center for Law Enforcement Technology at the University of Denver, and the Crime Prevention Analysis Lab at the School of Criminology at Simon Fraser University in British Columbia, Canada, and the vendors of GIS software have training centers in many cities. Sending a cadre of users for several days to be trained at such a center and sending more involved users for several weeks of specialized training, though costly, will greatly speed implementation. There are other learning resources available, such as various relevant books (mostly case studies of crime mapping or techniques for geo-spatial analysis) and an increasing number of general books on the topic of GIS and related technologies such as GPS.

Alternatively, the end result of GIS education and training (proficient GIS users) can be obtained by hiring a new employee possessing a background in this technology. Typically, new hires with degrees in geography, urban and regional planning and sometimes computer-aided design should have the requisite set of skills. Those graduating in computer science or MIS are unlikely to have such skills unless they went out of their way (i.e. beyond degree requirements) to obtain them. A small but increasing number of students in the field of criminal justice are getting at least some exposure to GIS in colleges and universities and these may be the most desirable potential *GIS analysts* for a law enforcement agency. Hiring a GIS analyst poses various problems for police departments. In most parts of the United States competition is fierce for qualified employees and turnover is rapid with qualified internally developed GIS professionals moving to greener pastures at frequent intervals. This process is greatly facilitated by the numerous GIS employment-related web sites and job postings on the Internet. Therefore, it behooves police agencies to offer competitive salaries, and in the case of sworn officers that develop an expertise in crime mapping and analysis, to promote them at a far faster rate than for typical line officers, even those adept in specialties such as "detective." The GIS expert that is still an "officer" after five years may well obtain a lieutenant's rank most easily by moving to another department, taking a wealth of community-specific knowledge and skills with him or her.

GIS implementation questions

There are a variety of questions that have been raised by potential and recent users of GIS that are important to address, as their answers can make for successful implementation and even help answer the more fundamental question of whether a GIS is really warranted in a specific organization. Presented below are some common GIS implementation-related questions along with some common sense answers.

Who needs GIS?

The most basic question related to adoption of GIS by law enforcement is which law enforcement agencies need GIS. Some might argue that any police department could benefit from GIS. This may be true, but is irrelevant, because for many smaller departments the costs and hassle of GIS outweigh the limited benefits. For many small departments what would be more useful than a GIS per se, would be a wall map showing the *current* streets and the relevant jurisdictional boundaries. This, coupled with a spreadsheet in which temporal data on various types of crimes and deployment of resources was maintained and which could be accessed to help generate charts and reports for each geographic subdivision (say by beat), would go a long way toward providing the same products that a GIS could provide in typical initial use. This wall map could incidentally be generated using a GIS, although more and more street maps based on the US Census Bureau's TIGER data (sometimes enhanced) are available over the Internet in a format that can be downloaded and easily printed at least as a small-scale map. Even the smallest police departments can benefit from an up-to-date plotted large-scale wall map. GIS use is not limited to megalopolises like London, England, New York City or Chicago, medium-sized cities such as Waco, Texas, and Spokane, Washington and even very small towns such as Broken Arrow, Oklahoma are embracing the technology. For example, the three-man police department in Cut and Shoot, Texas, uses such a map generated by the GIS analyst of a larger neighboring city (Figure 2.19).

It is probable that such an approach would work well in those cities with less than 10,000 persons and those departments with less than ten sworn officers. For departments with less than 100 sworn officers, a single copy of

Figure 2.19 An example of a small police department using GIS (indirectly) is the Police Department of Cut and Shoot, Texas (yes, that is the official town name!), which has a population of approximately 2,000, served by three full-time and five reserve officers. It uses an ArcView and ARC/INFO GIS generated wall map (provided by the GIS analyst of the neighboring city of Conroe, Texas) to map crimes (with pins) and delineate the growth of the city and its extra-territorial jurisdiction (which now encompasses an additional 8,000 people). The photo shows the Chief of Police and a patrol car in front of the town hall/police department building. *Photography by Mark Leipnik.*

a desk-top mapping program like MapInfo used by the crime analyst or the person responsible for maintenance of crime reports and preparing responses to uniform crime reporting (now the National Incident Based Reporting System or NIBRS) requirements or by an officer responsible for traffic-related issues would seem to be adequate. Some small departments, in wealthy cities such as Stonington, Connecticut, Cathedral City, California, and Scottsdale, Arizona, that have the resources, will often adopt a customized crime-mapping approach like the use of *CRIMEVIEW*. Larger cities will typically need multiple copies of a GIS, perhaps one in each precinct, preferably tied together by a network with a common base-map and an incident report database residing on a server. More sophisticated tools can be justified where a full-time GIS analyst or proficient crime analyst is available to take full advantage of the increased functionality.

Which types of law enforcement agencies can use GIS?

GIS use in law enforcement is concentrated in municipal police departments. However, an increasing number of sheriff's departments are using GIS. In fact, since sheriffs have responsibilities for larger spatial areas than typical municipal police departments, it might seem that they would be more likely to benefit from GIS than municipal police departments. However, with the population density and concomitant density of crime being lower in most of the areas that sheriff's departments patrol, GIS adoption by these agencies has been delayed. Nevertheless, many sheriff's departments in urban areas of California including Los Angeles, Ventura, and San Bernardino Counties (Figures 2.14 and 2.15), in suburban areas of Colorado, Florida (Figure 2.7), Oregon and Washington and in the greater Washington, D.C, metropolitan area, which includes parts of Maryland and Virginia, are using GIS. Most such departments are relatively urban departments. The exceptions include Bernadillo County, New Mexico and Nueces County, Texas.

Several states are using GIS either as a tool to support local law enforcement as in Delaware where the State Police are active GIS users or as in Illinois and New Jersey where State Troopers are using GIS and AVL systems to pinpoint fatal accidents, map and analyze locations of drunk driving arrests, drug interdiction stops and other incidents on the State Highways and other roads they patrol. Collected data is also being used to address the contentious issue of the possible existence of racial profiling in traffic stops.

Some organizations involved in community supervision of parolees have used GIS, but the results have been disappointing. Community supervision applications of GIS are lagging policing ones, perhaps because the location of prior offenders is generally something that these entities do not really wish to track, control or be held responsible for. Unless incentives can be created that make it more politically feasible to track where prior offenders are residing, working, or re-offending it seems unlikely that community supervision will be as fruitful an applications area as policing. Examples of community supervision applications include work in Cuyahoga County, Ohio, and in Wisconsin. The State Board of Pardons and Paroles is undertaking tracking registered sex offenders with GIS in Texas, as are several other states. Analysis of this data has indicated such interesting situations as a dozen registered sex offenders residing in the same Houston, Texas apartment house or the presence of a large state-sponsored "half-way-house" for pedophiles and other sex offenders located nearly adjacent to a children's day care center in Austin, Texas.

The full potential of GIS is only slowly being realized at the national level in the United States. There are examples of special projects by the US Drug Enforcement Administration, Bureau of Alcohol Tobacco and Firearms, the US Secret Service, Coast Guard and other agencies mostly aimed at drug interdiction or tracking money laundering. Use of GIS in counter-terrorism by the US, CIA, and Canadian Military Intelligence is increasing, but these uses are quasi-military and in any event are not being publicized. There is also the occasional use of GIS by the FBI Behavioral Sciences Division in profiling serial killers. The US Border Patrol's San Diego sector office is actively using GIS to delineate areas of illegal entry and also to target drug interdiction efforts. One outstanding example of use of GIS by the US National Guard Bureau is presented later in this book.

Outside the United States, GIS is widely used by law enforcement in Canada and the United Kingdom. The Royal Canadian Mounted Police are GIS users as are the municipal police of Vancouver, Edmonton, and Winnipeg (Rossmo 1995). In the United Kingdom GIS is used at the national, regional, and local levels of law enforcement. Specifically, it is being used by the Regional West Midlands Police in Birmingham, by the metropolitan Liverpool Constabulary and by local municipal (council) police in Harrow, a suburb of London. Material on law enforcement use of GIS in Great Britain by the South Yorkshire Regional Police is presented later in this book. Other countries that are employing GIS in law enforcement include Australia where the State Police of New South Wales are among the most active users, New Zealand, Belgium, The Netherlands, and the Czech Republic. The Munich Germany Police are leading users of the technology and are capable of generating daily crime maps and even are recording traffic infractions, a level of detail that is exceptional. The German Federal Police (BKA) are using GIS to create flow maps of the movement of illicit drugs into and through transshipment points in Europe to destination cities in Germany. In most instances, the GIS is being used primarily by national or regional policing agencies; this is in contrast to the United States where local police departments are the most active users.

What types of GIS generated products are most useful in law enforcement?

The answer to this question depends on the perspective of the GIS user. For a patrol officer or sergeant, either a paper map and report of crime incidents in the recent past (for their beat) or a map and attribute data delivered to a mobile data terminal are the most useful products. The difference between these two products is that the mobile data terminal (MDT) or field-use lap-top solution allows the officer to enter incident reports into a report template, and perhaps capture the location of an incident with GPS or query the digital map to extract the appropriate address and possibly even pull up internal building plans in a siege situation, all while on the scene. The downside of the availability of all this "real-time" information in the field, is that such an approach is only recently feasible and requires an investment of something like $15,000 per vehicle for MDT technology as well as a sophisticated digital radio transmission network. The paper map, on the other hand, requires a $199 printer (as well as software, data entry, and other resources required in either case). However, the rapid advance of wireless digital communication including location-based services and the integration of GPS and cell phone triangulation into personal digital assistant (PDA) devices offers the prospect that police officers can not only locate themselves in relation to Starbuck's coffee outlets (and donut shops) but will be able to use commercial and consumer technology to interact with GIS in the field.

For the crime analyst, the ability to interactively view, query, and analyze spatial and temporal aspects of a crime incident, study a series of related crimes or the general pattern of crime in a community is of greatest utility. The creativeness of the analysts is supported by a host of spatial and statistical analysis functions available in programs like ArcView and ARCGIS and its crime mapping extension or MapInfo Professional and in customized GIS crime mapping packages like CRIMEVIEW or the CrimeStat program from Ned Levine & Associates. Generating ellipses indicating "hot spots" of crime on top of street centerlines and beat boundaries is quite useful. Addition of features such as schools, liquor stores, or housing projects to this analysis adds greater predictive power. Alternatively, creating buffer zones around streetlights showing degree of illumination in relationship to nighttime crimes can be valuable in analysis and prevention of auto accidents. An analyst may also be able to visualize the relationship between bicycle paths and green belts and suburban burglaries from analysis of digital aerial photography (DAP) displayed in a GIS. DAP along with locations of attacks by serial rapists on schoolgirls in Detroit, Michigan helped an analyst create a geographic profile of one or more serial offenders helping elucidate patterns such as "hunting," "pouncing," and "scavenging." By understanding these behavior patterns and the spatial and temporal "profile" of an offender, an analyst may be able to predict where the rapist (or killer) may strike next and help design suitable surveillance and decoy efforts. These are all examples of the kinds of analysis being performed by crime analysts using GIS.

At the command level, decision makers may actually benefit from the simplest output from the GIS: color-coded (chloropleth) crime density or graduated dot maps of crime locations/rates and areas of increases and decreases (*exception reporting*) delineated for the jurisdictions for which they have responsibility. Community outreach efforts can also benefit from such products, especially if they can focus on the types of crime of relevance to the audience in question. Specifically, in a presentation to a neighborhood association, a detailed map of the area showing recent property crimes would be most useful. An example of a product designed for dissemination to the general public might be a "map" showing locations of certain registered sexual offenders, generalized to a given block, as mandated by laws ("*Megan's laws*") in various US states, requiring public notification of these locations (City of Redlands 1999).

What are the confidentiality and misuse issues?

This last example of a GIS product raises another difficult question. What about the confidentiality and accuracy of crime data stored in a GIS and portrayed on products such as maps? Data stored in a police department's GIS is theoretically as secure as in the days when all data was stored on paper or on a mainframe computer. In reality, this is not the case. By assimilating masses of data and displaying it in a comprehensible format in a variety of maps and other products, GIS now allows geo-spatial data to be disseminated far more easily and widely, particularly via the Internet. This can lead to a whole host of problems. For example, would data on the name, address, etc. of fraud victims be something to make public, particularly over the Internet, or would this information be of more help in identifying potential targets of opportunity to lurking swindlers than it would be to law enforcement or the public?

Even more sensitive data is stored in many systems, such as information about sexual assault victims, addresses, and phone numbers of witnesses and people with protective orders or the addresses of registered sex offenders. Generalizing this data by limiting information about locations to only the block on which the person resides, or omitting names, does not solve all privacy-related issues by any means. What about inaccurate data? Successful lawsuits have been initiated against police departments that released outdated but specific locational data for a sexual predator. This was found to have resulted in harassment and vandalism of the "innocent" residents of the offender's former home (Paisner 1999).

Attitudes about release of possibly sensitive data vary widely. Police in College Station, Texas have released maps of the residence locations of registered sexual offenders to the news media and these have been displayed on the local nightly television news. This same department presented a poster at the Crime Mapping Research Conference in Dallas, in December 2001, containing the photographs of registered sex offenders along names and (hopefully) current residence locations. In contrast, police agencies in other locations consider crime maps to be highly confidential. So, for example, the National Police Agency in Japan would be unwilling to release maps of actual sexual assault or other violent crime locations, even for a city the size of Tokyo, although derived products such as crime density maps, which cannot be associated with a specific location, are made available to outside groups (Yamamoto 2001).

Another example of sensitive data is provided by maps showing gang territory. These can be generated by creating hot spots around crimes with known gang associations as in Redlands and Salinas, California, and Chicago, Illinois, or even by mapping the distribution of gang-related graffiti as in San Antonio, Texas. These maps may be of internal utility but there may be negative public relations consequences from releasing maps portraying an area as being "the territory" of the "Black Disciples," the "Latin Kings," the "40 Ounce Crew" or the "Aryan Nations." This information might imply to some of the public that the gang's activities in this area were being accepted or even endorsed by the powers-that-be.

As more and more entities get access to GIS, it becomes increasingly likely that geo-spatial crime data will find its way into the hands of those outside law enforcement who will put it to various uses ranging from setting insurance rates, to locating businesses, to selecting schools for their children. For example, several major insurance companies have begun to use crime data from national sources to set insurance rates for various types of property. They use GIS and a customized decision support system to adjust citywide crime numbers reported to the FBI down to census-tract level subdivisions with independently obtained demographic data (TIGER data) and their own geo-coded internal claims information. Availability of far more geographically specific crime data would accelerate this trend. Likewise, websites that claim to compare the overall crime rates of cities (only for major US cities at present) would be able to offer more and more specific data, presumably for a fee. People and companies will use this information when making relocation decisions. Such analysis is likely to further marginalize crime-ridden areas by retarding redevelopment efforts, although such information could conceivably indicate neighborhoods where actual crime rates were lower than perceived crime rates.

An example of the public relations problems that can result from inadequate analysis or misinterpretation of crime-mapping data by persons outside law enforcement is illustrated by the situation occurring recently in one of the ten largest metropolitan areas in the United States. In that city, a map of overall crime rates per capita delineated by census-tract, was generated by the news media based on police department data and published in a leading daily paper. The analysis indicated very high overall crime rates for a premier downtown business and shopping district. This result was an artifact of the very low "resident" population enumerated by the 1990 census compared to the vast numbers of shoppers and office workers present in the area on a daily basis. Only a tiny proportion of these "visitors" were actually victims or perpetrators of crimes while in the district, and even then the charge was typically petty larceny. The downtown merchants association and by extension the mayor and city council did not appreciate that data from a police department GIS had helped generate this result, albeit a statistical "fluke." Thus, as the availability of crime data in a geo-coded format expands and the ease of use of GIS increases, more and more opportunities for misuse or reuse of the data and technology will arise. One can easily anticipate opponents of a proposed public housing project or residential treatment center doing their own crime density or "hot spot" analysis of similar existing projects within a city. One can imagine that a mayor or city councilperson or city housing authority representative would not appreciate such an analysis and its related maps seeing the light of day. This is not withstanding published work sponsored by the US Department of Housing and Urban Development which purports to demonstrate lower crime rates around public housing projects (Hyatt 1999).

How can inter-agency cooperation in GIS development be achieved?

The answer is that the best way to take advantage of all the benefits that the GIS has to offer is for police departments to work cooperatively. Other local and regional government departments and law enforcement agencies may be able to provide financial resources, data, or expert staff to ease the learning-curve, speed development time and reduce the financial and man-power burden of GIS development. There are many examples of governmental entities likely to be using GIS. City or county community development and comprehensive planning departments can provide data on streets and zoning and land-use issues. City or county public works, engineering or road and bridge departments can also provide street data. City or county tax appraisal districts are often early adopters of GIS, and they have street and property ownership and improvement data. In many areas, regional councils of government (COGs) are coordinating GIS development and they may even have funds for local police departments to get software and training. COGs may be able to provide regional data on streets and jurisdictional boundaries. Some state agencies such as State Departments of Transportation have road data (although often only for state and county roads and then usually in a Intergraph CAD format). State natural resources and environmental agencies often have data and expertise with GIS. At the US Federal Government level, the US Census Bureau has built and maintains the TIGER data set which is the most popular one used in geo-coding incidents, although enhanced versions of this data-set compatible with specific GIS packages are the medium of choice despite their higher cost (Myers 1992).

Other law enforcement agencies may be able to share data and the burden of GIS development. Offenders do not usually concern themselves with determining in which jurisdictions they are committing crimes. Numerous examples of crime issues that cross jurisdictional boundaries exist. Most common are crimes in suburban areas committed by inner-city offenders, but conversely some problems, like methamphetamine labs, are concentrated in rural areas but attract customers from nearby cities (Rengert 1998). A GIS whose street data ends at the city limit is likely to miss the overall pattern of crime affecting a city. Examples exist of neighboring cities (Bryan and College Station, Texas) with incompatible GIS approaches (ESRI versus Intergraph) using an appraisal district's GIS to map a crime spree spanning both cities. A better approach would be for all law enforcement agencies at both the municipal and county level to cooperate on development of a common base-map and share data about crimes occurring throughout the region. Examples of such cooperation include work among city and county law enforcement in the greater Baltimore, Maryland, area and joint GIS efforts by the Corpus Christi Police Department (Texas) and the Sheriff's Office of the enclosing county (Nueces County, Texas) (Canter 1997). The cases of Pinell as County Florida and Delaware presented later in the book exemplify the value of such interjurisdictional cooperation. A variant of the shared approach would be for the largest or most experienced entity in a region to assist smaller entities in GIS development. This makes a great deal of sense in cases where small independent enclaves exist in or near large urban areas or where a single city is disproportionately larger than any other in a region. Alternatively, a county or regional government could assume responsibility for mapping all crimes in the entire county or region; this is an approach being adopted in Great Britain and New South Wales, Australia. A similar approach is illustrated by the cooperation in crime mapping and analysis provided by the Illinois State Police to the beleaguered municipal police of Springfield, Illinois, a smaller downstate city that has been overtaken by big city drug and gang-related problems. The location of the State Police headquarters in Springfield (which is the State capital) facilitated this cooperation (Bitner 1999).

Ideally, other city or county agencies may be able to take responsibility for maintaining the base-map and particularly for incorporating new streets and may be able to provide other types of data such as location of features like city owned property, fire stations, parks, and schools in the community. Unfortunately, relatively few municipal and county GIS databases are structured to perform geo-coding, so their base-map of streets, although usually more accurate than even enhanced TIGER data, may prove of limited assistance unless reconfigured. In turn, the police department can provide other municipal departments valuable data. For example traffic accident locations could improve the transportation planning and road, bridge and streetlight maintenance function. Some communities, which have taken community planning and policing to heart, would also view crime data as a useful adjunct to the comprehensive planning and community redevelopment process. Thus, areas with hot spots for various crimes associated with abandoned buildings, vacant lots, single resident occupancy (SRO) hotels, etc. might be specifically targeted for increased building safety compliance monitoring, redevelopment and civic improvement related activities. For example, in Charlotte, North Carolina, the Police Department's GIS identified a vacant lot downtown as being a "hot spot" for various assaults, robberies, narcotics, and vice offenses. Network access to the cadastral GIS maintained by the tax assessors office, determined the property's ownership. Rapid abatement was effected because the owner was determined to be the City's redevelopment authority! This agency at the urging of the police chief installed fences, lighting and other improvements that mitigated crime associated with the property, pending its eventual reuse (Novicki 1998).

These and other examples of implementation-related issues and the successes that can be achieved through diligent development of a comprehensive crime mapping and analysis capability within law enforcement agencies are provided throughout the remainder of the book.

References

Bitner, L. (1999) Conversation with Mark Leipnik.

Block, R. L. (1995) Geo-coding crime incidents using the 1990 TIGER file: The Chicago example. In: C. R. Block, M. Dabdoub, and S. Fregly (eds) *Crime Analysis Through Computer Mapping*, Washington, DC: Police Executive Research Forum.

Block, R. L. (1997) Conversation with Mark Leipnik.

Campbell, J. (1998) *Map Use and Analysis*. 3rd edn, Dubuque, IA: William Brown & Co.

Canter, P. R. (1997) Geographic information systems and crime analysis in Baltimore County, Maryland. In: D. Weisburd and J. T. McEwen (eds) *Crime Mapping and Prevention*, Monsey, NJ: Criminal Justice Press.

Canter, P. R. (1998) Baltimore County's autodialer system. In: N. LaVigne and J. Wartell (eds) *Crime mapping case studies: Successes in the field*, Washington, DC: Police Executive Research Forum.

City of Redlands (1999) Transforming Community Policing for the 21st Century: Risk Focused Policing. Redlands California Police Department, Redlands, CA.

Dent, B. (1999) *Cartography: Thematic Map Design*. 5th edn, Boston: WCB/McGraw Hill.

Hyatt, R. A. and H. R. Holzman (1999) *Guidebook for Measuring Crime in Public Housing with Geographic Information Systems*. Washington, DS: U.S. Department of Housing and Urban Development.

Korte, G. (1992) *The GIS Book*. 2nd edn, Santa Fe, New Mexico: On Word Press.

Leipnik, Mark, K. Kemp, and H. Loaiciga (1993) Implementation of GIS for water resources planning and management. *Journal of Water Resources Planning and Management* 119: 2.

McGuire, P. (1998) Conversation with Mark Leipnik.

Myers, D. (1992) *Analysis with Local Census Data*. Boston, MA: Academic Press.

Novicki, D. (1998) Conversation with Mark Leipnik.

Paisner, S. (1999) Exposed: On-line registries of Sex Offenders may do More Harm Than Good. *The Washington Post*, 21 February.

Rengert, G. (1998) *The Geography of Illegal Drugs*. Boulder, CO: Westview Press.

Rich, T. (1996) *The Chicago Police Department's information collection for automated mapping (ICAM) Program*. Washington, DC: U.S. Department of Justice National Institute of Justice.

Rossmo, D. K. (1995) Overview: Multivariate spatial profiles as a tool in crime investigation. In: C. R. Block, M. Dabdoub, and S. Fregly (eds) *Crime Analysis Through Computer Mapping*, Washington, DC: Police Executive Research Forum.

Sorensen, S. (1997) SMART mapping for law enforcement settings: Integrating GIS and GPS for dynamic near-real time applications and analysis. In: D. Weisburd and J. T. McEwen (eds) *Crime Mapping and Crime Prevention*, Monsey, NY: Criminal Justice Press.

US Census Bureau (1997) *TIGER/Line Files Technical Documentation*. Washington, DC: United States Census Bureau.

Yamamoto, Tetsuya (2001) Conversation with Mark Leipnik.

Focus Box

GIS use in Waco, Texas

Mark R. Leipnik, Donald P. Albert, Dennis Kidwell, and Albert Mellis

Waco is a moderate-sized central Texas city, far removed from other major metropolitan areas. It is centered in a farming region and features some light manufacturing and a well-known university (Baylor). The population of the city is approximately 109,200. The city has a substantial African-American population (23 percent) and a growing Hispanic community (16 percent). Crime rates are unexceptional. The crime-related issue that most people associate with Waco, the Branch Davidian confrontation, occurred a dozen miles outside of town, and the Waco Police Department had little or no role in it. The Waco Police Department is made up of 201 sworn officers, 74 civilians, and 400 "citizens on patrol."

The Waco Police Department has been using GIS for approximately ten years and currently is routinely generating crime analysis bulletins for its patrol officers. In general, these bulletins are geared toward property crimes and feature a cover page addressing current crime issues (for example, highlighting the likelihood of burglaries near the university during spring break). Other pages contain citywide and zoomed-in maps that display incident locations along the boundaries of districts and individual beats. A final section contains the list of actual incidents that relate to the incidents symbolized on the map (Figure FB2.1, Table FB2. 1, and FigureFB 2.2 for a representative sample of a bulletin related to residential burglaries). The GIS is used by the officer responsible for both crime analysis and crime reporting to generate such routine reports, which are disseminated by the low-cost method of copying and distribution as paper flyers delivered directly to the hands of supervisors and officers. Alternatively, the GIS can address issues of current interest such as the zoomed-in map of assaults (Figure FB 2.3). Close-up maps for particular crimes and the associated tabular data can be easily created for all or part of Waco and for any chosen incident type or period.

The Waco Police Department's GIS system also has the capability of performing several types of spatial and statistical analysis. Specifically, crime density maps can be generated (see Figure FB2.4). The system stores seven years of prior incident records that can be used to highlight areas where crimes are likely to occur based on predictive statistical models and these

CRIME ANALYSIS BULLETIN

DATE:03-06-00 **BULLETIN:00-002**

CRIME: RESIDENTIAL BURGLARY JANUARY 01 - March 4, 2000

TIME MOST BURGLARIES OCCURRED:	1000-1400 & 2200-0000
DAY MOST BURGLARIES OCCURRED:	Tuesday, Wednesday, Saturday & Monday
METHOD OF ENTRY MOST OFTEN USED:	FRONT & REAR DOOR AND REAR WINDOW
PROPERTY MOST OFTEN STOLEN:	TV, VCR, CASH, HEATERS, JEWELRY, FIREARMS & LAWN TOOLS
RESIDENTIAL BURGLARIES JANUARY 1 TO MARCH 4, 2000:	139
AREAS TO WATCH:	Beat 220 Gorman to Herring from N 15th to N 27th

It's Spring Break Time, Watch The Apartments and Houses Around Baylor University.

Dennis Kidwell, Sergeant
Crime Analyst
Criminal Investigations Divisions
Waco Police Department

Figure FB2.1

Figure FB2.1–2.5 Examples of crime analysis and mapping from Waco Texas Police Department. All figures courtesy of Alberto Mellis, Chief, Waco, Texas Police Department.

areas can be compared with the location of current incidents (Figure FB2.5). This relatively low-cost system, which utilizes primarily commercial off-the-shelf GIS technology (MapInfo), has helped Waco improve its delivery of services to the community.

Table FB2.1 Residential burglaries January 1–March 4, 2000; suspect information

Case	*Date*	*DOW*	*Time*	*Address*	*Suspect 1*					*Suspect 2*					*LIC No*	*Suspect vehicle*				
					Race1	*Sex1*	*Age1*	*HGT1*	*WGT1*	*Race2*	*Sex2*	*Age2*	*HGT2*	*WGT2*		*Year*	*Make*	*Model*	*Style*	*Color1*
00000591	01/02/2000	SU	21:50	Confidential data	H	M	0	0	0	H	M	18	502	180	Confidential data	97	CHEV	CAPRICE	4DR	WHI
00000843	01/03/2000	MO	11:00		W	M	27	0	0			0	0	0		0				
00001101	01/05/2000	WE	10:00		B	M	20	510	140	B	M	20	510	140		0	CHEVY	CHEVETTE		BRO
00002105	01/10/2000	MO	11:15		W	M	21	510	140			0	0	0		0				
00003514	01/17/2000	MO	00:50		B	F	15	500	130	B	M	20	507	160		0				
00003772	01/18/2000	TU	09:45		B	M	0	511	180			0	0	0		0				
00004222	01/20/2000	TH	14:00		B	M	0	600	160	B	M	17	511	145		85	OLDS	DELTA 88	4DR	BRO
00004257	01/20/2000	TH	:		H	M	11	0	0	H	M	11	0	0		0				
00004482	01/21/2000	FR	22:15		W	M	15	0	0	W	M	15	0	0		0				
00004822	01/23/2000	SU	04:00		W	M	0	508	190			0	0	0		0				
00006641	02/02/2000	WE	09:45		B	M	20	0	0	B	M	20	0	0		0				
00007674	02/07/2000	MO	00:00		B	M	19	0	0			0	0	0		0				
00007935	02/08/2000	TU	07:45		H	M	25	600	190	H	F	34	505	140		0				
00008346	02/10/2000	TH	06:44		X	M	0	506	0			0	0	0		0				
00008869	02/12/2000	SA	:		H	M	20	0	0	H	F	20	0	0		0	DODGE		MINIVAN	RED
00009515	02/13/2000	SU	13:00		B	M	25	0	0			0	0	0		0				
00009874	02/08/2000	TU	:		W	M	20	600	0			0	0	0		0				
00009910	02/16/2000	WE	17:20		B	M	18	510	175			0	0	0		0				
00010927	02/21/2000	MO	15:00		W	M	0	0	0	W	M	0	0	0		0				
00012264	02/26/2000	SA	:		H	M	21	0	0	H	M	21	0	0		0	CADI		4DR	GLD
00012312	02/26/2000	SA	17:30		H	M	16	0	0			0	0	0		0				
00013005	02/29/2000	TU	14:50		B	F	29	506	300			0	0	0		0				
00013601	02/15/2000	TU	08:00		B	M	20	0	510	B	M	0	600	0		0				
000136	03/03/2000	FR	12:00		W	F	18	508	200	H	F	16	503	115		0				

Residential Burglaries January 1–March 4, 2000

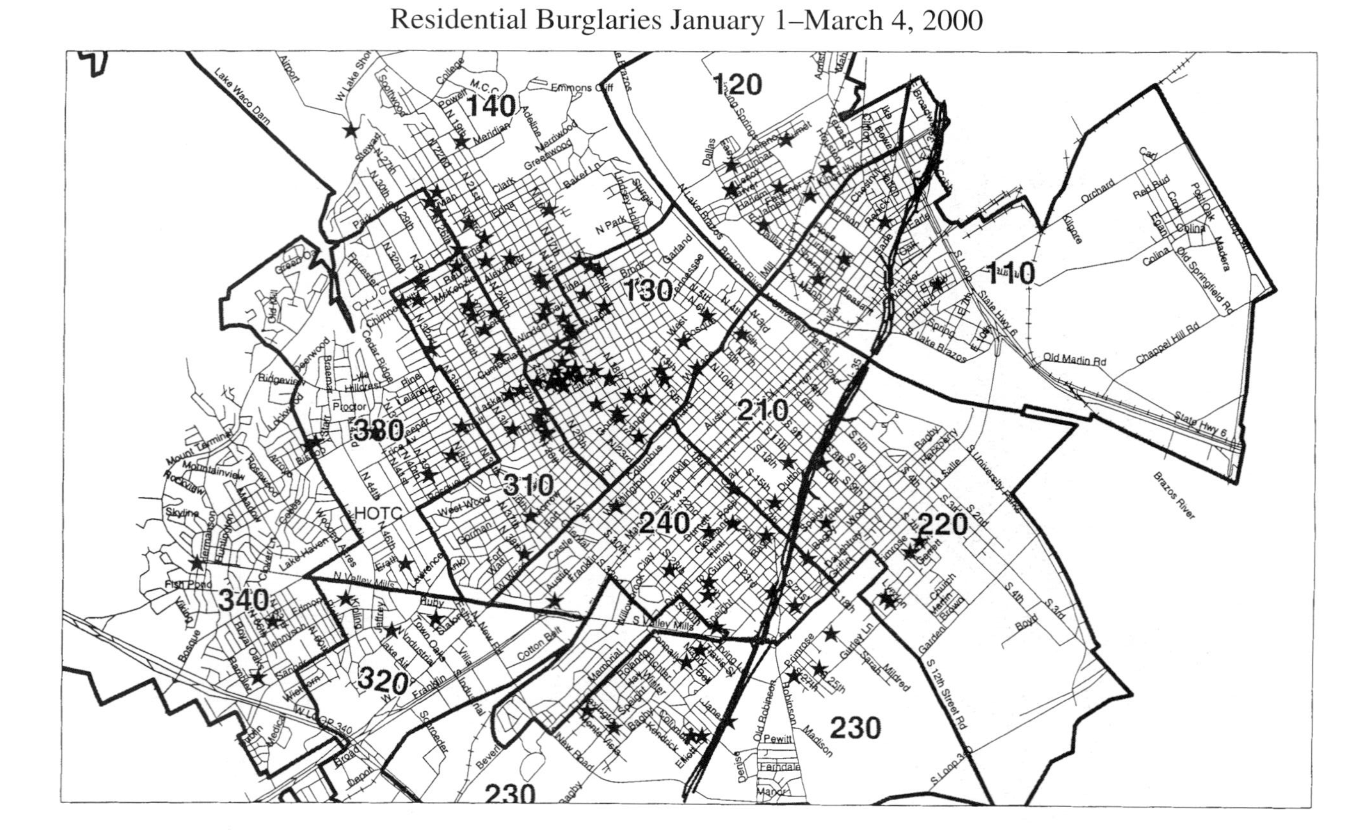

Figure FB2.2

Assaults June 01–15, 2000

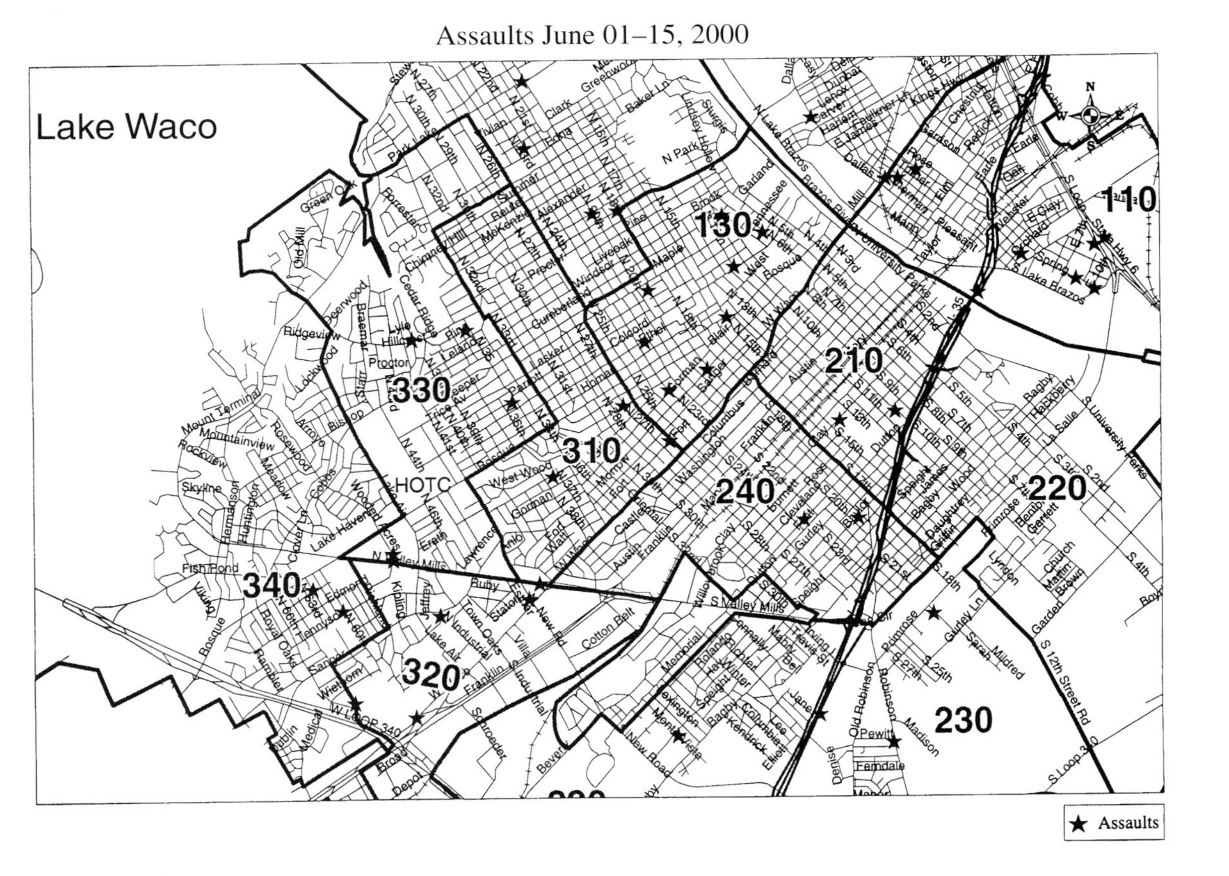

Figure FB2.3

Residential Burglaries January 01–June 12, 1999

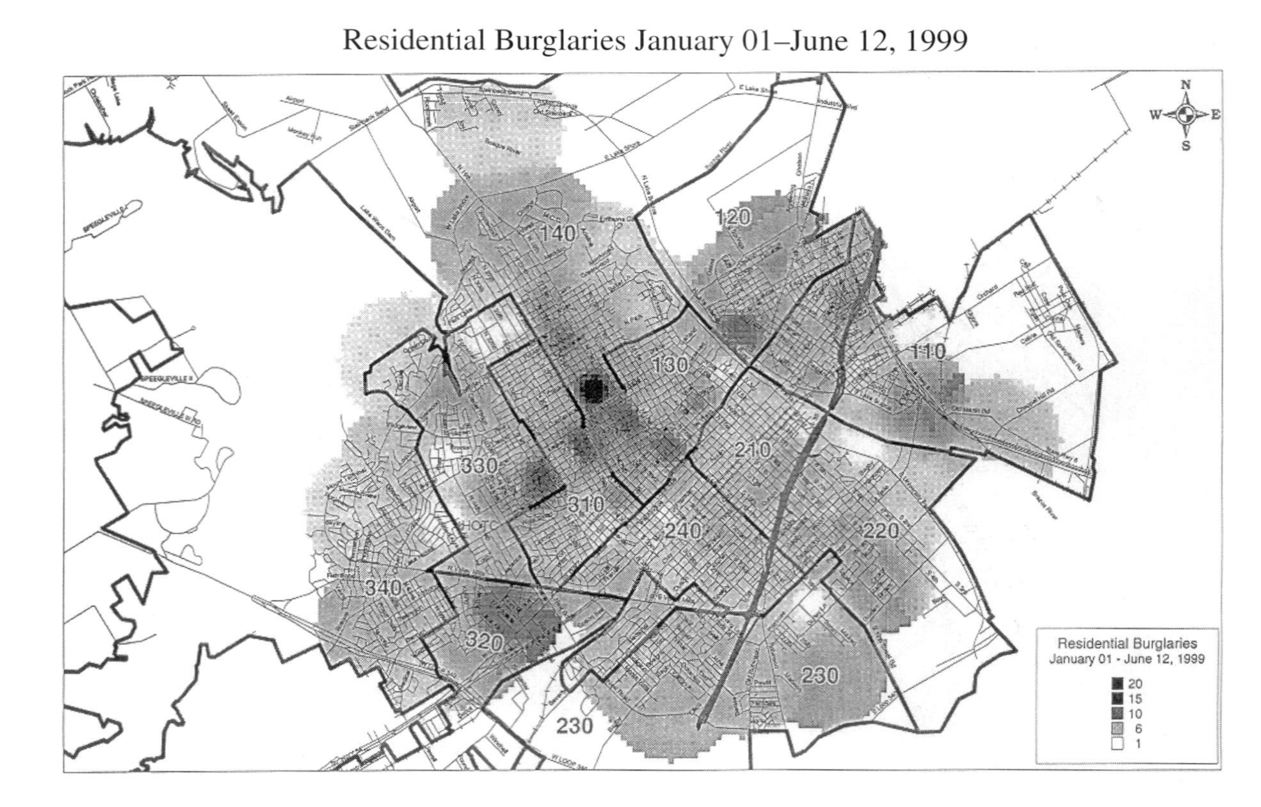

Figure FB2.4

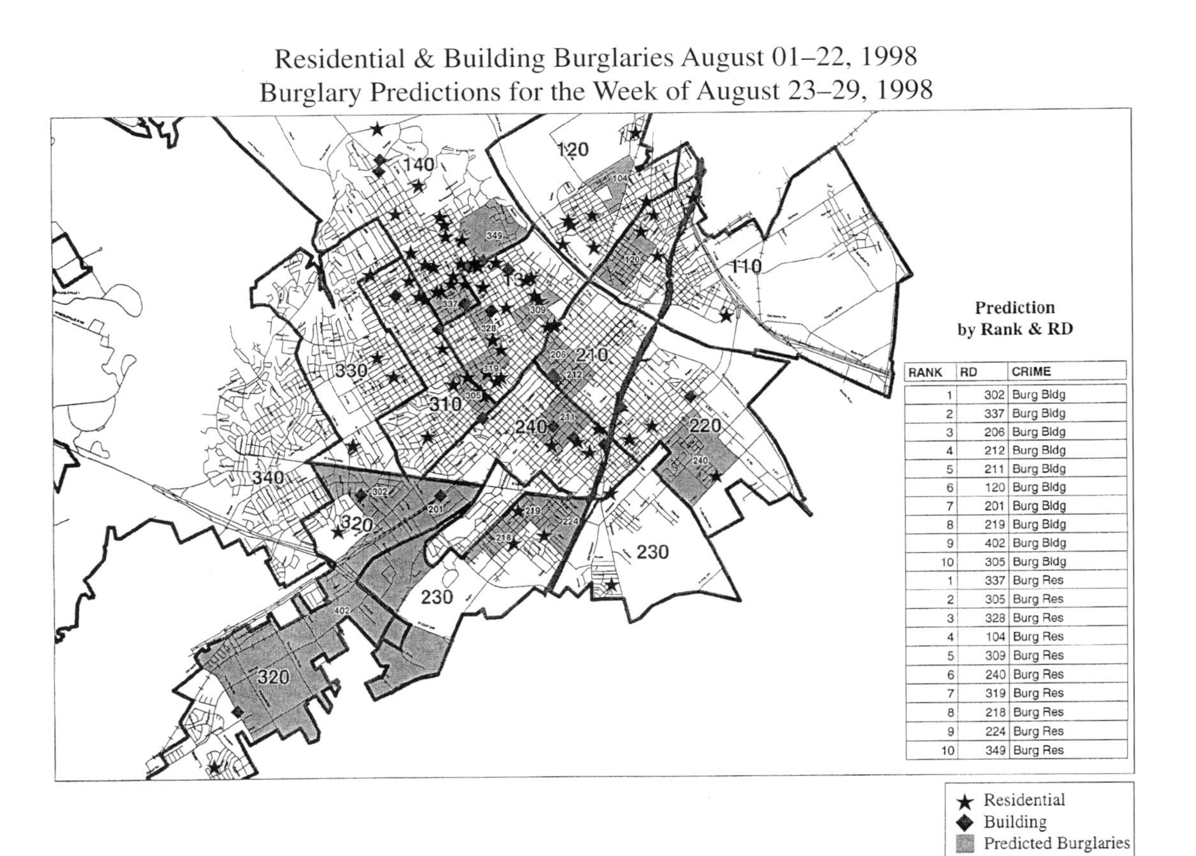

RANK	RD	CRIME
1	302	Burg Bldg
2	337	Burg Bldg
3	206	Burg Bldg
4	212	Burg Bldg
5	211	Burg Bldg
6	120	Burg Bldg
7	201	Burg Bldg
8	219	Burg Bldg
9	402	Burg Bldg
10	305	Burg Bldg
1	337	Burg Res
2	305	Burg Res
3	328	Burg Res
4	104	Burg Res
5	309	Burg Res
6	240	Burg Res
7	319	Burg Res
8	218	Burg Res
9	224	Burg Res
10	349	Burg Res

Figure FB2.5

3 Mapping in Mayberry

Major issues in the implementation of GIS in small and rural law enforcement agencies

Derek J. Paulsen

The use of geographic information systems (GIS) has increasingly become recognized within law enforcement as a highly effective means of enhancing crime analysis through the analysis of crime patterns, evaluation of crime trends, and assistance in strategic planning and problem solving. Due to increased publicity about its benefits, interest in acquiring and using GIS within law enforcement has grown exponentially. Recent research suggests that the majority of police departments feel that crime mapping would be a valuable tool for their department in dealing with crime (Mamalian *et al.* 1998). While a few years ago crime mapping might have seemed cutting-edge and out of reach, the declining costs of computer hardware and software have placed GIS firmly within the grasp of many law enforcement agencies. In accordance with this, the use of GIS has increased over the past few years, with approximately 16 percent of the more than 13,500 local police agencies using GIS to some extent in 1999 (Hickman 2001).

While these numbers appear to paint a rosy picture for the future of GIS within law enforcement, a closer inspection into the use of GIS within policing points out some disturbing disparities in its use. Most importantly, the use of GIS in law enforcement is increasingly becoming limited to large police departments that make up only a fraction of the total number of law enforcement agencies nationally (Hickman 2001). Specifically, 90 percent of the eighty-six agencies that serve populations over 250,000 and approximately 60 percent of the 537 agencies serving populations between fifty and 250,000 are currently using GIS (Hickman 2001). In contrast, only approximately 14 percent of the nearly 13,000 departments serving populations less than 50,000 are currently using GIS (Hickman 2001). Moreover, the number of small departments currently using GIS actually declined between 1997 and 1999 (Hickman 2001). This is all the more striking given that small departments comprise approximately 90 percent of all law enforcement agencies nationally. While numerous articles have been written discussing implementation of GIS in law enforcement, currently none have dealt specifically with the issues facing small and rural law enforcement agencies wishing to implement GIS (Garson and Vann 2001; Hughes 2000; Mamalian *et al.* 1998; Vincent 2000). Although articles that deal generally with implementation of GIS are instructive and beneficial, the failure to address the specific issues faced by small and rural agencies may actually have contributed to its limited use within these agencies. Accordingly, the purpose of this chapter is to explore some of the major issues specific to implementation of GIS in small and rural law enforcement agencies and to provide a simple guide for implementation within these agencies. Specifically, this chapter will discuss issues relating to needs assessments as well as software, hardware, and training options.

Planning stage

After an agency makes the determination that it wishes to acquire a GIS, the first step is to plan the implementation of GIS (Garson and Vann 2001). This planning stage has several key aspects that are essential to a successful implementation of GIS within an agency, the first of which is to enlist the support of key police managers (Garson and Vann 2001). Prior research has shown that one of the keys to successful implementation of GIS is the philosophy and attitude of police managers toward GIS (Cope *et al.* 1998; Garson and Vann 2001; Wellar 1993). Implementation of GIS is much more likely to be successful if those police managers in key decision-making positions within an agency are convinced of its importance and effectiveness in achieving departmental goals. While in large departments this may imply convincing many members of middle and upper management of the benefits of GIS, in smaller departments this usually implies convincing the Chief of its effectiveness. Although numerically this stage is much easier for small and rural departments, many small and rural agencies suffer from an inability to find sound information concerning GIS capabilities and benefits. Various sources are available for educating police managers about the use and benefits of GIS, including neighboring departments currently using GIS, GIS vendors, GIS consultants, universities, law enforcement periodicals such as *Police Chief* and government agencies such as the Crime Mapping Research Center. This key support is essential in overcoming officer-resistance that is almost certain to develop with the implementation of GIS. This officer resistance can be crippling to the implementation process and can take several different forms such as fears of new technology, belief that GIS is simply a fad, and general resistance to change within policing (Harries 1999). Police managers who are well educated and informed as to what can be done with GIS and how it can help a department achieve its goals, are more likely to lead a department through implementation problems that may arise.

After securing support from key police managers involved in the decision-making process, the second step in the planning stage revolves around conducting a needs assessment (Garson and Vann 2001). The goal of a needs assessment is to attempt to gather information about what an agency both wants and needs out of a GIS system. This information is then used to determine the specific acquisition needs of the department in terms of software and hardware as well as training and data needs. Information gathering for a needs assessment is usually conducted in both formal and informal sessions that involve as many people within the department as possible. While this can be a lengthy, expensive, and difficult procedure for large departments, information gathering is generally much easier for small and rural agencies because of their smaller size. Successful methods of information gathering include formal and informal surveys, open discussions, lunch meetings, and informal discussions between the Chief and officers.

While a needs assessment is vitally important in GIS implementation in a department of any size, it is all the more important in small and rural departments. Because of the significant costs associated with setting up a GIS, it is essential that small and rural agencies fully determine their exact requirements for the GIS in order to acquire a system that fully meets their needs, but yet fits within their limited budget. In implementing a GIS, several specific questions should be addressed, including: who will be responsible for managing the GIS, how will GIS be integrated into the department, who will the end-users be, data issues, and what will be mapped (Garson and Vann 2001). It is in addressing these specific questions that small and rural agencies differ most from larger departments. Specifically, the amount and type of incidents that occur in small and rural jurisdictions coupled with budget and manpower differences impact greatly the GIS needs and capabilities of small and rural agencies.

In conducting a needs assessment, the first issue that small and rural agencies must address is determining *who* will manage the GIS once it is acquired. Management of GIS is an issue that should not be taken lightly, as who maintains the GIS and how much time they have to do so are often key elements in whether GIS implementation is a success or failure. In large departments, GIS management is usually handled in one of two manners, hiring of new GIS personnel or providing current data management personnel with training in GIS. Importantly, these options are not always available to small and rural agencies. Except in the rare case of a wealthy small agency, the hiring of new personnel, expressly for GIS management, is not possible due to budget limitations. While grant funds are available to assist in hiring situations such as these, most grants are for limited time periods, may only cover acquisition of technology and in many cases require an agency to pay at least half of the new personnel's salary. Thus, because of budget limitations, most small and rural agencies will have to rely on current personnel for GIS management. However, small and rural agencies usually have fewer available staff than larger agencies, with a far greater percentage of total personnel being sworn officers than in large departments. In many small and rural agencies, data management is handled part-time, either by sworn officers or civilian personnel such as clerical personnel. Thus, in determining beforehand who will be responsible for GIS management, small and rural agencies must also assess the impact these additional duties will have on that individual's other duties within the agency. Importantly, if a sworn officer is chosen for data management, this may result in a severe reduction in their law enforcement duties. Another factor for small and rural agencies to consider in assigning GIS management duties, is the training required by personnel for its effective operation and use. Learning how to use GIS effectively and efficiently can be a daunting task for even those with extensive computer experience. In most cases, the best option is to either seek volunteers who are interested in learning GIS or to appoint the individual within the agency with the most computer knowledge and aptitude. Importantly, these issues are vitally important in making a determination about the type of GIS system that is best suited for an agency. Those agencies where time and training limitations of personnel are major issues would be well advised to acquire a GIS system that is easy to learn and may have limited functionality. Conversely, those agencies where GIS management issues are not a problem may be better able to handle acquiring a sophisticated GIS system. While every small and rural agency must make its own determination concerning GIS selection and management, those that take these issues into consideration will find GIS implementation a smoother process.

The second issue small and rural agencies must address is determining how GIS will be integrated into the department. A major issue in the acquisition of any new equipment or service by a law enforcement agency is determining how best to integrate it within the agency's structure so that its full potential and benefits can be realized (Garson and Vann 2001). In implementing GIS the major issue for agencies is determining how the information that a GIS provides will be integrated into the decision-making process. One of the main benefits of a GIS is that it provides law enforcement with a new method of analyzing the locations of crimes, calls for service and problem areas, and this analysis can be used to develop new patrol, problem solving, and crime prevention strategies (Harries 1999). Currently, GIS is commonly used to assist in the decision-making process in a wide array of areas, such as the creation of new patrol beats, directed patrol locations, crime prevention programs, and other strategies that make policing efforts more effective and efficient (Harries 1999). While large departments may be able to easily use the information provided by GIS analysis to implement changes, small and rural agencies may not be able to implement the strategies that are suggested by GIS analysis.

In discussing the use of GIS to deal with "hot spots" of crime, one chief of a small police department related two major barriers to its effective use (B. Post, personal communication, 3 April 2001). First, many small and rural agencies have no training in how to respond effectively to these "hot spot" situations. For many small and rural agencies, strategies such as directed patrol, crime prevention through environmental design, and problem-oriented policing are programs they may have heard of, but have no experience with. Thus, acquiring training for police managers and decision-makers is essential, for without training, designing, and implementing these and other strategies may prove daunting for many small and rural agencies (Garson and Vann 2001). A second issue that was raised in using GIS in decision-making revolves around the need for manpower in carrying out these strategies. While a large department may be able to respond to a crime "hot spot" by using extra patrol in the affected area, small and rural agencies are limited in the manpower they can devote to an area. Thus, even if many small and rural agencies know *how* to respond to an emerging "hot spot," because of manpower limitations they may not be able to respond effectively. As one chief stated in discussing "hot spot" responses, "I would only be able to respond with an officer or two and even that might require overtime pay"(B. Post, personal communication, 3 April 2001). In integrating GIS within an agency, small and rural agencies must recognize the limitations they will have in using GIS effectively, and plan beforehand how best they can use its capabilities to meet their goals and needs. A sound plan concerning the role GIS will play in decision-making and problem solving is essential for its capabilities to be fully taken advantage of. Moreover, deciding how GIS will be integrated into decision-making will assist in the acquisition of a GIS system that will best fit the needs of the department. Without sound planning in this area, many departments may purchase a GIS system with more capabilities than they need, wasting precious money that could have better been spent elsewhere.

A third issue that must be addressed in a needs assessment concerns the potential end-users of GIS products. Importantly, the issue of potential end-users is closely related to integration of GIS into a department. How well a GIS is integrated within a department will determine to some degree who the end-users will be. Of all the different issues to be addressed in a needs assessment, none is arguably more important than the determination of end-users of the GIS. A complete list of end-users and their needs will determine the range of end products that a GIS must produce and thus the functionality of the GIS. Those agencies that foresee having many different groups of end-users will need to have more flexibility and capability in their GIS system than those with a more limited variety of end-users. Interestingly, this is one area of GIS use where small and rural agencies do not differ greatly from larger law enforcement agencies. Traditionally, end-users of GIS analysis in law enforcement are limited to four main groups, each with different needs. The first group of end-users is line officers. Line officers traditionally use GIS analysis to assist them in duties related to patrol and thus require maps that are simple to understand, yet detailed enough to provide an understanding of the conditions in their patrol area such as their beat. The second group of end-users is police managers. Police managers use GIS analysis primarily for decision-making and strategic planning purposes and thus need maps that provide an understanding of trends and patterns on a smaller scale (covering a larger area in less detail such as the entire city) than those of line officers. In larger departments crime analysts and task force members may be active GIS users. In small and rural departments these specialties are typically absent. The third traditional group of end-users is other government officials such as city council members. These other government officials use GIS analysis to help them understand crime related issues and needs of law enforcement and thus require maps that can show how law enforcement needs are being met or neglected. The final group of traditional end-users is community members. Community members use these maps to inform themselves of crime related issues in their communities and thus require maps that are simple and easily understandable. Departments need to determine in advance both the potential end-users and the products that they will need to develop.

A fourth issue to be addressed by small and rural agencies, and one that is often overlooked in determining what type of GIS system an agency needs, is data issues (Garson and Vann 2001). Data issues are extremely important in GIS implementation and consist of three main issues, an agency's current data management system, base maps and geo-coding of incidents and "other" data to be used. The most important of these three issues concerns an agency's current data management system. This issue is of great importance because agencies must be sure that they are capable of exporting data from their records management system (RMS) into their GIS to prevent inefficient use of GIS. There are numerous cases of law enforcement agencies that have acquired an expensive GIS system only to have it gather dust because they found out after purchase that they have no way of exporting incident data from their RMS into their GIS. This is a particularly acute problem in small and rural agencies where RMS systems tend to be more archaic than in larger departments. Thus, to avoid this potential problem, an agency should discuss data management issues with its GIS vendor before purchase of a system or integrate GIS acquisition into an overall modernization of the crime records management system.

A second data issue deals with base maps and geo-coding issues that small and rural agencies will encounter after acquiring a GIS. The most common method of inputting data into a GIS is through a process called geo-coding, which involves taking incident addresses and matching them to a base map containing all street addresses for an area. While this can be a relatively easy process for large urban areas, geo-coding is often a much trickier proposition for small and rural agencies where good base maps are not always available. Agencies must determine if good base maps are available for their jurisdiction, and if not, they need to make plans to either purchase one, hire a contractor to create one or cooperate with a local or regional government in developing one. In addition, data collection may have to be altered in many small and rural agencies in order to make sure that official street addresses, rather than common place names, are used to delineate where an incident occurred. Finally, agencies must determine what "other" types of data they want to analyze besides crime incidents in order to make plans for acquiring this data. Importantly, many different sources of good contextual cultural data are freely available for use and can be acquired easily from other government agencies, the US Census Bureau or GIS vendors.

The final issue to be addressed in a needs assessment is *what* an agency is going to map. This stage of a needs assessment is vitally important because it will help determine the end products available to the various end-users. What types of issues that an agency will map is impacted mostly by the nature and extent of crime incidents within a jurisdiction (Harries 1999). In determining what will be mapped, an agency should make a list of the most important problems that they face and how mapping can assist in those problems. It is here that the greatest differences exist between large urban law enforcement agencies, and small and rural law enforcement agencies. While police agencies that serve jurisdictions with populations under 25,000 account for approximately 90 percent of all law enforcement agencies, in the US they account for only approximately 16 percent of all reported crime, and only 11 percent of all serious violent and property crime (Pastore and Maguire 2000). In general, because of the differences that exist in the nature and extent of crime in small and rural jurisdictions, these agencies may not need GIS systems that are as sophisticated as those necessary in jurisdictions with larger populations.

A comparison of two cities of different sizes in North Carolina illustrates this point. In the town of Boone, a university town with approximately 14,000 year-round residents, the most serious problems facing the police are vandalism, auto accidents and alcohol related incidents such as driving while intoxicated (DWI) arrests. In contrast, in the city of Charlotte, with well over 500,000 residents, crime problems are similar to all major cities, including, drug sales, violent crime, and police resource allocation issues. Because of the differences in the nature and extent of crime incidents in these two cities, what they map and the GIS system required to meet those mapping needs will be very different. In Boone, the mapping of auto accidents, vandalism locations and DWIs can be accomplished with digital "pin maps" and this will require only a simple GIS. In contrast, the crime problems in Charlotte require advanced analysis techniques such as crime forecasting for resource allocation, geographic profiling for investigations and crime density change maps for line officers, all requiring a far more advanced GIS system than Boone. As with other needs assessment issues, each small and rural department must make its own determination of what it will map based on the nature and extent of crime incidents and the needs of the end-users within its jurisdiction. Carefully delineating these issues in the planning stage will help prevent small and rural agencies from making poor decisions when it comes to acquiring a GIS system.

Acquisition phase

After a department has carefully conducted a needs assessment, the next step is to acquire a GIS system. Importantly, it is in this stage that many small and rural agencies experience their largest problems. The main cause of these acquisition problems revolves around attempting to match the needs and requirements determined in the needs assessment with a GIS system that will fit a small or rural agency's budget. One of the main problems that small and rural agencies have in attempting to make this match of needs and budget is a lack of understanding of the options available to them. Often small and rural agencies purchase systems that are more expensive and have more functionality than they require to meet their limited needs. As a result, these systems often end up being used sparingly and resentment grows towards GIS because of the substantial expense when compared to the limited benefits derived. The GIS options available to small and rural agencies can be classified into five different categories: in-house development of GIS, use of other government agencies, obtaining university assistance, cooperation with a regional mapping initiative, and contracting for the assistance of private service providers (See Table 3.1). This section is designed to act as a guide to these different options. It discusses the different benefits, problems, and basic hardware requirements associated with each option. Small and rural agencies should use this section, and its associated table, as a reference to guide them in making general choices about how to acquire GIS analysis capabilities.

The first crime mapping option available to departments is to acquire a stand-alone desktop GIS system such as ESRI's ArcView program, MapInfo's MapInfo, Microsoft's Mappoint, or Integraph's GeoMedia to name a few. These desktop systems are notable for their flexibility and functionality, providing agencies with the widest range of analysis capabilities and the ability to upgrade. In addition, several of these desktop systems benefit from a wealth of available software extensions that can customize the system specifically for crime analysis. Other benefits include dedicated technical support from the vendor, integration capabilities with other government agencies, and a network of current users within law enforcement who can provide both practical and technical assistance to new users. However, along with these many benefits comes a wealth of potential problems for small and rural agencies. The biggest problem with these desktop systems is their cost. In addition to the initial purchase cost, there are associated training costs, service contract costs, and costs of eventual upgrades and extensions. For many small and rural agencies, acquiring a desktop system is simply not a cost-effective decision as it provides far more functionality than they need at a cost higher than they can afford. Minimum hard-ware requirements for these desktop systems are not usually very onerous, usually being accommodated by a PC compatible computer running a 166 Mhz Pentium or higher chip with at least 64 Mb RAM, a good color monitor and a quality color printer. Because of the costs associated with these systems, they are usually best suited to agencies that desire full functionality and timely data analysis.

The second option available to small and rural agencies is to enlist the assistance of other government agencies in their community such as

Table 3.1 GIS options for small and rural police agencies

System type	*Cost*	*Analysis capabilities*	*Associated benefits*	*Associated problems*	*Recommended for Agencies that*
In-House within police department	Moderate to expensive	Advanced	• Flexibility and functionality • Available software extensions • Dedicated technical support • Integration with other agencies • Network of existing users	• Cost of software • Service contract cost • Extensive training required • Cost of upgrades and extensions • More functionality than many departments need	• Need functionality • Need/want advanced spatial analysis ability • Have multiple end users • Have complex end products • Cost and training not an issue
Other government agency takes lead	Inexpensive	Basic	• No purchase of software, hardware, or training required • Integration with other government agencies • Time saving	• Analysis conducted when conven for other agency, not always when convenient for police agency • Security and privacy of incident • Analysis with no crime training	• Desire entry level analysis only • Have limited funds or manpower • Have GIS management and training problems • Have limited end users and mapping products
University takes lead	Inexpensive	Basic	• Dependent upon partnership • Integration of crime data with contextual data • GIS knowledge and experience • Minimal initial investment	• Availability of a University • Timeliness of analysis • Security and privacy of incident	• Desire simple analysis • Have limited funds or manpower • Have limited end users and mapping products • Have GIS management and training problems
Regional crime mapping program shares lead	Inexpensive	Basic	• Cost savings on software, hardware, and training • Crime viewed in regional context • No security or privacy issues	• Logistics of collaboration with other agencies • Data agreement issues • Cost sharing issues • Individual agency responsibilities • Trust issues • Timeliness of data	• Desire a regional perspective on crime issues • Have limited funds of manpower • Have limited end users and mapping products • Have GIS management and training problems
Application service provider takes lead	Moderate to expensive (dependent on services desired)	Basic to advanced	• Cost savings on software, hardware, and training • Customized mapping applications and services • Expert crime analysis and problem solving assistance • Dedicated technical assistance • Maps on demand	• *Potential* security and privacy issues concerning incident data • Outsourcing of data • Availability of timely data	• Desire a wide array of analysis capabilities and expertise, while minimizing software, hardware, and training costs • Outsourcing of GIS is not an issue

a planning department, zoning board, tax assessor, or other agency currently using GIS. This option allows small and rural agencies to avoid many of the costs associated with the acquisition of GIS software, hardware, and training, yet still conduct basic GIS analysis of incidents. Importantly, the costs associated with acquisition of software, hardware, and training are often listed as the biggest barriers to departments conducting GIS analysis (Mamalian *et al.* 1998). Other benefits include the integration of crime data with data from other government agencies and the saving of time by not conducting the analysis in-house. However, conducting GIS analysis in this manner also engenders several different problems, chief among them being the timeliness of analysis. When crime mapping is outsourced, the analysis is usually conducted when it is convenient for the other agency and not when the police needs it. This can have severe implications in its utility in decision-making and full integration within a department. Other problems include security and privacy of incident data and the lack of training in the analysis of crime by the analyst at the "sister" agency. Because of these limitations, this option is best suited for agencies that want entry level analysis, where GIS management and training are major issues and where there are a limited number of end-users with limited mapping needs.

The third option available for small and rural agencies is to enlist the assistance of a local university in the analysis of crimes. This option can take several different forms ranging from complete outsourcing of GIS analysis, as with the aforementioned option, to the university providing manpower (in the form of interns) for in-house GIS development. The benefits of this option differ depending on the assistance received by the agency, but generally include lower costs associated with software and training as well as expert assistance with GIS development. This option allows small and rural agencies to benefit from those who have extensive experience, prior knowledge, and experience with GIS, without requiring a heavy investment of resources from the agency. Other benefits include integration of crime incidents with other contextual data and reduced workload on existing officers. Problems associated with this option are dependent on the degree of assistance provided, but chief among them is the availability of a university to provide assistance. Many small and rural agencies are not located in areas where assistance from a university is practical. Other problems associated with this option include the timeliness of the analysis and privacy and security of incident data. Because of the limitations of this type of GIS analysis, this option is best suited for agencies that want simple analysis as in the previous option.

The fourth option for small and rural agencies is to become involved in a regional crime mapping program. These programs are similar to both university and government agency assistance in that GIS analysis is essentially outsourced. However, in regional crime mapping programs it is another law enforcement agency that conducts the analysis rather than a civilian agency. In addition to the obvious cost savings for software, hardware, and training, there is the added benefit of viewing local crime incidents in the context of crime incidents from other jurisdictions. This regional crime analysis allows agencies to better understand the crime problem in their area by seeing crime incidents beyond their own jurisdictional borders. In addition, as opposed to other outsourcing options, there are no real security or privacy issues as all analysis is conducted by law enforcement agencies. Problems associated with this option are those that are endemic to all law enforcement collaborations, namely logistics. In order for regional crime mapping programs to work smoothly they require a data agreement by all participating agencies, in which all agencies will have a common base-map and tables with common database designs. Importantly, this is often very difficult for diverse agencies with different data gathering methods to agree upon. Other problems include determining cost sharing issues, selection of which agency will be responsible for the actual analysis duties and general trust issues involved in police collaborations. Ideally, the data would be accessed from a server either over a dedicated wide-area network or via the Internet with password based access. A regional mapping program is best for those agencies that desire a regional perspective on crime, where GIS management and training and costs are an issue.

The last option for small and rural agencies is to have GIS development and/or analysis conducted through a private application service provider such as TYR Systems or the Omega Group, Inc., of San Diego California. Application service providers, or ASPs, are companies that provide complete customized GIS services for a low monthly price, usually based on the services desired and the amount of incidents within a jurisdiction. The services provided vary greatly from simple pin maps delivered via fax, to intranet system hosting allowing officers to create their own maps, and more advanced analysis techniques and internet crime map hosting. Besides the savings in software, hardware and training costs, other benefits of ASPs include expert crime analysis and problem solving assistance, maps on demand, and technical assistance. Like other outsourcing options, the biggest problem associated with ASPs are logistical issues and the potential security or privacy issues. However, unlike other outsourcing options, an ASP is a business that focuses on providing GIS services; thus, they are far less likely to have security or privacy problems than other outsourcing options. An ASP is suitable for a wide array of agencies, ranging from those wanting to experiment with GIS to those with advanced analysis requirements.

Training

Another major issue that an agency will deal with in implementing GIS is to determine where and how to receive training in the use of GIS. Training is a crucial element of GIS implementation and an aspect of GIS that tends to be ignored (Garson and Vann 2001). GIS training is necessary not only for GIS analysts, but also for those individuals who will be end-users and consumers of GIS information. As discussed earlier, one of the problems that many agencies may encounter in attempting to integrate GIS into their decision-making process is the lack of training in how to respond to problems that GIS identifies. In addition to initial training, agencies must also be sure to maintain their knowledge base through ongoing training and where appropriate longer term education. Importantly, training can be received in many different ways, including through government sponsored training, regional and national crime analyst associations, vendor sponsored training, university classes, and in-house or self-paced training (Garson and Vann 2001). Government sponsored training is available from organizations such as the Crime Mapping Research Center of the National Institute of Justice, the Carolinas Institute for Community Policing, or the National Center for Technology. These government sponsored training centers provide cost effective training that is specific to either ArcView or MapInfo, the two most popular desktop GIS systems. One problem associated with this form of training are the limited locations and times that are currently available for training. A second training option is to receive training from regional or national crime analysis associations. As with government sponsored training, these associations provide good cost effective training in Arcview and MapInfo formats from actual crime analysts. Similar to government sponsored training, training from crime analyst associations is also limited in the times and locations. A third method of GIS training is to receive training from the GIS system vendor. While these training sessions are more expensive than others discussed so far, they are by far the most intensive and complete. The fourth GIS training method is to receive training at community colleges or universities. The many universities offer basic training in GIS, with an increasing number offering classes specifically dealing with crime mapping. While university offerings are excellent for basic GIS training, they are often lengthy in duration and present personnel scheduling problems. The final method of training is in-house or self-paced training. This option is best for those who are capable of learning well on their own. Selection of which option or combination of option is best will vary with the specific needs and resources of the agency. More importantly than *where* individuals receive training is that at least analysts and managers receive some GIS training.

Conclusion

Although many small and rural agencies desire to use GIS and can benefit greatly from its capabilities, difficulties in the implementation of GIS have prevented most of them from doing so. While following the advice in this chapter will not guarantee a successful implementation of GIS, in a small or rural department, an understanding of the issues and problems presented here will make implementation success far more likely. In implementing GIS small and rural agencies are encouraged to explore fully all of the issues they have concerning integration of GIS, end-users, GIS management, data issues and what will be mapped in order to determine their exact GIS needs. Furthermore, agencies should be cognizant of the issues that they will face in terms of training and maintenance, as these are essential to successful use of GIS and continued satisfaction with GIS products. Finally, agencies must be sure to persevere in their implementation efforts, as the benefits of GIS analysis are well worth the struggle of implementation.

References

Cope, J. *et al.* (1998) Five step method to regional crime analysis. *The Police Chief* 65: 46–49.

Garson, G. D. and I. B. Vann (2001) Management strategies for implementing GIS in municipal law enforcement. Paper presented at the meeting of the American Society for Public Administration, Newark, N.J.

Harries, K. (1999) *Mapping Crime: Principle and Practice.* Washington, D.C.: U.S. Department of Justice.

Hickman, M. J. (2001) Computers and information systems in local police departments, 1990–1999. *The Police Chief* 67(1): 50–56.

Hughes, K. (2000) Implementing a GIS application: Lessons learned in a law enforcement environment. *Crime Mapping News* 2(1): 1–5.

Mamalian, C. A., N. G. La Vigne *et al.* (1998) *The Use of Computerized Crime Mapping by Law Enforcement: Survey Results.* Washington, D.C.: U.S. Department of Justice, Office of Justice Programs, National Institute of Justice.

Pastore, A. L. and K. Maguire (eds) (2000) Sourcebook of criminal justice statistics [Online]. Available: http://www.albany.edu/sourcebook/ [4/10/01].

Vincent, K. (2000) Implementing crime mapping. *Crime Mapping News* 2(1): 6–8.

Wellar, B. (1993) Key institutional and organizational factors affecting GIS/LIS strategies. *Computers, Environment & Urban Systems* 17(3): 201–212.

Focus Box

Crime maps, Overland Park Police Department, Kansas

Gerry Tallman

Overland Park, Kansas, has a population of 151,734. The Overland Park Police Department (215 sworn officers) publishes a wide range of crime maps that help both officers and citizens identify potential problem areas. The following is a description of some of their standard crime mapping products.

Weekly property crime map

Previously titled the Weekly Crime Map, it is produced every Friday. This map (actually six maps on a 2′ × 4′ poster) shows the location of all residential burglaries for the previous 7, 30, and 90 days and auto burglaries, auto thefts, and commercial burglaries for the previous 30 days (Figure FB 3.1). Graphs depict data on premise type, point of entry, entry tool, day of week, and time of day. Copies of this map are hung in each Division and in public access areas of the Department. Additional copies are provided to the City Manager/City Council and published in the local newspaper.

Weekly persons crime map

Produced every Friday, it is similar to the Weekly Property Crime Map and contains six maps showing the location of all robberies, purse snatches, rapes, window peepers, and lewd and lascivious incidents for the previous 90 days. It also provides the personal descriptors and vehicle descriptions of all registered and known sex offenders in the city and plots their home addresses on a separate map. This map is for internal use only and is not placed in public access areas.

Monthly traffic safety map

Currently under development, this product will contain six maps depicting fatality/injury accidents, non-injury accidents, juvenile accidents, speeding violations, following too closely, and traffic complaint areas. Graphs will show the top twenty-five traffic accident locations and display various causal factors, time of day, and day of week.

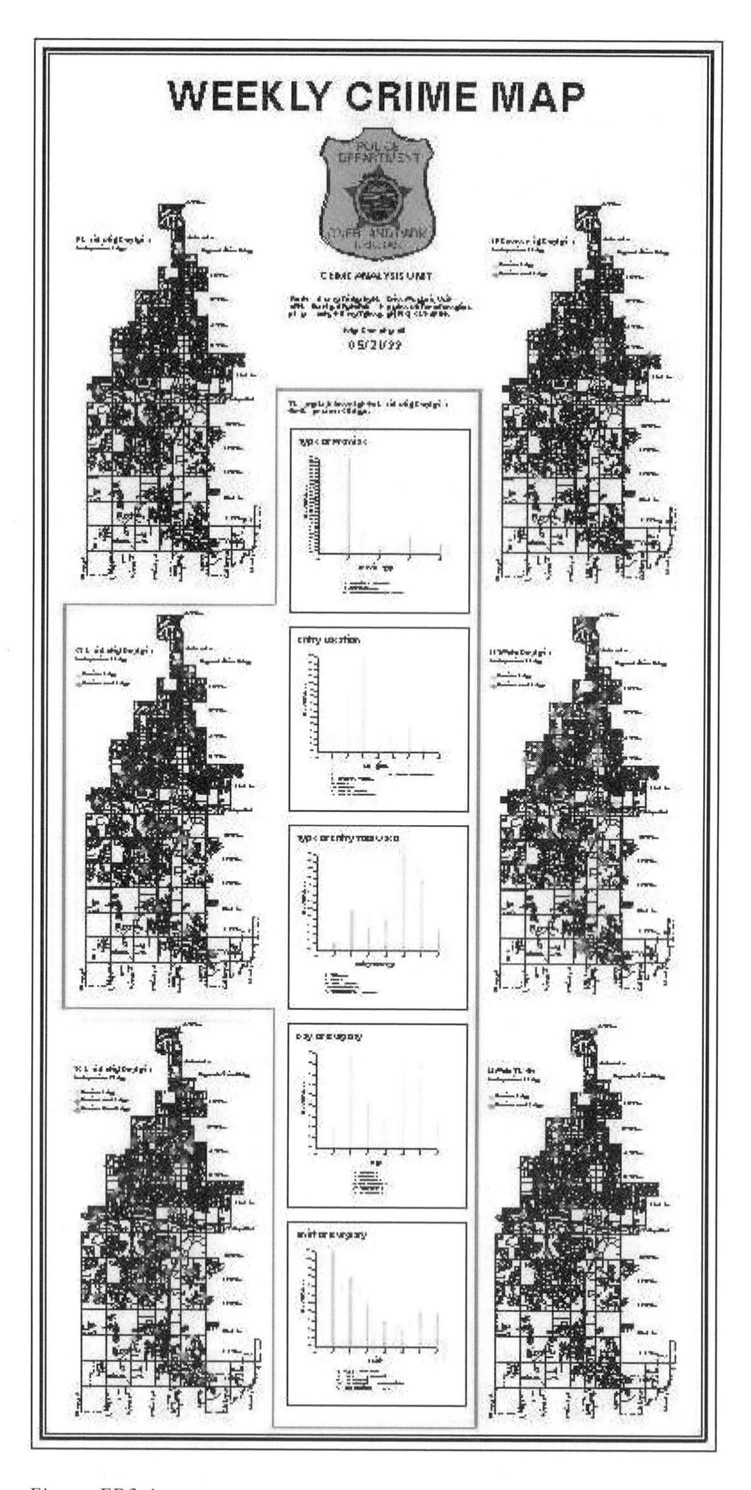

Figure FB3.1

4 Designing a database for law enforcement agencies

Erika Poulsen

The uses of GIS in law enforcement are expanding. A GIS is only as good as the data it contains. An agency can spend large amounts of money on acquiring hardware and software, but the most important component is the data. The mapping and analysis capabilities alone are not enough for a GIS to provide information for law enforcement agencies; a lot of data is needed, and the data should have a temporal component and must be able to be linked to other relevant community information (Block 1997). Agencies must have a clear idea of what the GIS will be used for in designing the database. It is important to have a database that will fit the agencies' needs, and deciding these needs first will reveal what type and detail of data is needed. In the end, a well-designed database should be able to incorporate data from different sources; link to external sources of data; and perform geographic, attribute, and/or temporal queries.

Database designs

Several systems have been created to analyze crime data spatially. The systems described here are all similar in that they incorporate data from a variety of sources, namely crime and community data, but each differs with relation to the scale of the area it services. The first system, GeoArchive, was developed to analyze crime and to assist law enforcement agencies at the community level. The spatial crime analysis system (SCAS) was designed for local police agencies. Finally, the regional crime analysis geographic information system (RCAGIS) handles data and conducts analysis on a regional level.

GeoArchive, developed by the Illinois Criminal Justice Information Authority, is a great example of a database that combines both community and crime data for the purpose of crime prevention (Block 1997). GeoArchive is a set of guidelines that law enforcement agencies can use to set up a database using data from different agencies to study crime at a community or neighborhood level. This is more of a methodological design rather than an existing software template, which instructs agencies on how to build a customized database. The resulting database uses both crime and community data to study crime, assist with investigations, and for problem solving. It can therefore serve as "an information foundation for community policing" (Block 1997, 28). As GeoArchives are created for the community level of study, it is important that community and crime data are available at the address level. The benefits for this level of data is that it can help law enforcement agencies identify "community problems" but the database does not have to be designed solely around one specific neighborhood. Several GeoArchives can be built for surrounding communities using the same design protocols so that data may be shared between the adjacent communities (Block 1997). GeoArchive is designed so that the data can be easily incorporated into a spatial analysis program, or a GIS. The advantage of using the GeoArchive methodology as opposed to the other databases structures mentioned below is that the resulting database will not be tied to one statistical analysis program.

The second database highlighted here is the SCAS developed by the United States Department of Justice Criminal Division's GIS staff and the INDUS Corporation. This database works for municipal level data. The SCAS has been implemented in the Montgomery County Police Department (MCPD) in Maryland. The MCPD manages five different databases which are maintained in different offices, and the goal was to build a central database (USDOJ 1997a). The needs for this database were to focus on spatial analysis, data sharing between local agencies, and to generate reports. This database was designed to have a consistent data structure for the five different offices in order to be able to share data easily and to create one centralized database that could combine the data from the offices.

Unlike GeoArchive, this database was created using Microsoft Access, and the resulting database design was then bundled with SCAS as a package product for other law enforcement agencies to use along with technical documentation (USDOJ 1997b). The analysis software package is a customized version of ArcView, and the agency must have the Spatial Analyst extension in order to operate such functions as crime density analysis. The database interfaces with the customized ArcView application through an open database connectivity (ODBC) interface (Nulph *et al.* 1997).

The RCAGIS was created by the US Department of Justice, Criminal Division and the INDUS Corporation. Initially, they had created the SCAS for local and medium-sized law enforcement agencies to analyze crime and then moved toward a regional approach with RCAGIS (Burka 1999). The RCAGIS was created to provide information about crimes that occur inside and outside jurisdictional boundary lines and to analyze crime that moves out of or into different localities. For example, if a City X wants to locate clusters of crime and only uses the data within the city limits for the analysis, an edge-effect phenomenon will occur. If there is a small cluster of incidents within the center of City X and a relatively larger cluster of incidents in the adjacent Town Y with a small portion of the incidents falling into City X, it would appear as though there is a strong cluster in the center of

City X and isolated events occurring at the edge adjacent to Town Y. The RCAGIS is designed to avoid this problem. The database was also designed to provide a resource to analyze serial crimes that cross jurisdictional boundaries (Burka 1999).

Each cooperating agency collects the data and uploads it to a centralized database and has the ability to download updated data from the database provided by other agencies. The data-sharing abilities are dependent on how frequently each cooperating agency provides an updated data set to the centralized database. Other agencies using these data must be aware of when these databases were updated, so that they may obtain the most recent data (Burka 1999).

Unlike SCAS, RCAGIS is a "stand alone" system, meaning that there are no additional software products to purchase in order to use the product. This would make the product much more affordable for agencies. The RCAGIS was created with Visual Basic 5.0, MapObjects, and CrimeStat (Burka 1999).

The RCAGIS can import data from different agencies that have differing data structures, coordinate systems, and projections and display the data layers as one seamless layer (Burka 1999). As an example, the RCAGIS can incorporate the street data layers from all the participating agencies, but these streets may not be in the same coordinate system. Instead of having all the street data layers showing up as several different data layers, the RCAGIS will display all the street layers as one file with a single geo-referencing system.

Each cooperating agency was required to provide certain uniform fields such as crime types, in order for the attribute data to be seamlessly mapped, but each agency could also provide additional fields that they deemed necessary. All of the crime data had to have a unique ID that could be used to link the data back to the originating agency's database to obtain further detailed information (Burka 1999). Just like SCAS, an ODBC structure was used to interface the data with the software.

Building a database

Conceptual design

The needs of the law enforcement agency must be defined in order to develop a database that will be useful. Once it has been established by an agency that a GIS will provide a new or assist an existing function, the mission of the GIS should be defined. This sounds simple enough; however, with the growing availability and accessibility of data and GIS software, it is easy to put together a database and quickly generate electronic pin maps of crime incidents. But these pin maps may not fulfill the agency's need for information. A conceptual plan for what information is needed is the first step in the development of the database.

A feasibility study should be conducted when a law enforcement agency decides to undertake GIS development. This study helps to establish the costs and benefits of the project. According to Yourdon (1989), studying the existing process may find problems in the existing system, such as

- variables not being collected from crime incidents
- data redundancy
- problems with data storage
- old police records being stored in paper format in an unsecured or non-fireproofed area and
- current computerized system reliability (out of date, or unable to handle current data).

A systems analysis should then be performed to define the current system that the agency uses to perform the tasks that are currently being performed and what new functionalities they require (Yourdon 1989). For example, if the GIS is being implemented for resources deployment, how are the police officers currently assigned their "beats?" Are the beats determined by time, space or some combination of both? Two work flow diagrams should be created: one to show visually how these processes are currently conducted and a second to display the system after it has been revised to incorporate a GIS.

The character of database will depend on the function of the agency and the level, and type of crime existing in the immediate and surrounding jurisdictions. Currently GIS databases are being used for deployment of resources, analysis of crime patterns, or in crime prevention programs. Each of these applications involves different types of spatial analysis and, to some degree, will have different data requirements. This is not to say that a database cannot be created that can fulfill each of these functions; but to ensure that the GIS will be useful, it is important to define what it will be used for and design the database around this definition. This definition is also important in that it will save valuable time and resources by collecting and importing only the necessary data. For instance, if an agency needs the GIS to assist with deployment of resources but also knows that a future crime prevention program will be installed, the knowledge will set the basis for determining what data will be needed for the GIS.

Another part of the conceptual plan is also to determine who the users are going to be. This will assist in determining whether or not to train existing personnel or hire additional persons trained in handling geographic information. While the decision for adding or training personnel should be conducted in the initial needs assessment for the GIS, during the conceptual phase of the database design, it may be concluded that a database administrator may be needed depending on the size of the agency and the functions and complexity of the database.

There are two methods for gathering information for the functions of the GIS. The first method is to approach different members of the agency establishing their functions within the agency, and finding out what they expect from the GIS. This is a good method to find potential uses for the GIS that may have been overlooked. The other method is to start at the top and work down. First to approach the head decision makers to find out what they expect from the GIS and then to work down through the agency to find out what administrators and officers want. This approach is beneficial in that the results can be compared to find out what are the best uses of the GIS for all persons within the agency. Conducting interviews with various levels of personnel from the management to the potential end-users will reveal how personnel in the agency envision that the GIS will automate functions and how the GIS may improve or effect the agency. For example, the top management sees the whole picture of the agency (the "forest") while the analyst works on a particular section of the agency (a "branch"). Consequently, knowing how each member sees how the processes of the agency works will provide a better view of how to build the system. During these interviews, the technical skill level of the users must also be established to determine training needs.

Another big task during the conceptual phase is locating existing and potential data sources. This can be done by interviewing various members of the agency, contacting vendors, talking to other agencies who have implemented similar databases, looking at current police reports, and by contacting professional societies and academia (Yourdon 1989).

A GIS database can house three types of information: geographic, attribute, and temporal characteristics (Openshaw 1995). For a law enforcement agency, geographic information could include crime incident data, political and natural boundaries, public infrastructure, and commercial and residential properties. According to Richard Block, the following data layers should also be considered for inclusion: land use, schools, community organizations, recreational facilities, emergency facilities, public health, and census data (Block 1997). Attribute data is the characteristics of the geographical features present in the GIS (DeMers 2000). Temporal data would include time, day, month and/or year the incident occurred; and temporal variations of components, such as holiday shopping, tourism season, school year, and any other fluctuations in the temporal rhythm of the geographic area that could have an impact on crime rates.

Geographical area

The definition of a study area is important to the design of the database (DeMers 2000). Determining the user of the data will assist in this decision. Will the database be used by multi-jurisdictional agencies? Is this database going to be used for micro or macro-scale studies? For example, does the agency need to study regional or statewide trends in crime? Or is the law enforcement agency concerned only with local crime incidents? By answering these questions, the spatial extent of the GIS will be revealed.

Attribute characteristics

An agency should establish which queries and analysis would be performed on the attribute data to assist in establishing which type of data should be imported into the database. Classifications may be designed so that selections may be performed for different characteristics of the crime incident: victim, target, suspect, or modus operandi. A numeric value could be assigned to crime incidents based on the severity of the crime; this code could then be used as a weight when analyzing crime patterns density.

When dealing with data from multiple agencies that share data, only key variables for the crime data might be entered into the shared database; departments may then independently share additional data about the crime incidents such as in the RCAGIS system. When working with multiple agencies, it is also important to note that the data are going to by very "dynamic" and will need to be updated frequently (Arvanitis *et al.* 2000).

Temporal

Time is an important factor in studying crime. Characteristics and patterns of crime will shift over time; therefore, an area may experience different crime issues over time (Vasiliev 1996). When implementing a prevention program in an area, a temporal study can be conducted to see how crime patterns adjust to this prevention program. Frequently, displacement of crime will be observed as a result of strategies such as "routine" patrols. Time is also an important component in studying trends in crime rates to find out if there has been an increase or decrease of crime within the study area. "Maps may need to be prepared on a weekly, monthly, quarterly, or annual basis to illustrate trends. Time-based maps may be further refined temporally by adding the shift dimension" (Harries 1999, 11). See the focus box in Chapter 3 for examples of temporal analysis of residential burglaries in Overland Park, Kansas.

In the conceptual design phase, temporal considerations must be considered. For example, is the GIS going to be used for deployment purposes? Does the agency want to schedule the deployment rotation based on temporal changes in crime? Such responsive short-term planning is needed to enable crime prevention methods to succeed (Gorr *et al.* 1999). Police agencies tend to manage police resources around shifts that may have more to do with personnel considerations than crime occurrences.

The time interval chosen in the database design will determine the precision with which the temporal aspects of crime behavior can be analyzed. Thus, if crime incidents are recorded by shift then hourly analysis of crime occurrence will not be feasible. Determining the exact time an incident like a residential burglary occurred can be problematic. The Overland Park Police Department uses the earliest hour a burglary could have occurred if uncertainty exists, so if a home is burgled over summer vacation the first hour of the vacation is taken as the time the incident happened. In contrast the Chicago Police Department uses the time the incident was discovered by the victim. Thus, 7 am Monday morning appears to be a peak time for motor vehicle theft. When observing criminal events, if an agency were to look at only monthly data, weekly, daily, even hourly fluctuations may become masked (Harries 1999).

Logical design

This section will discuss the methods with which to build the database. A structured method is recommended for the most efficient design. This will flag errors early on in the design process. This is especially useful when building a database that will gather large data sets from multiple agencies. The logical design includes determining what fields for data will be set up, how the data will be entered into the database, how these data tables will connect to each other if a relational database is employed, how the data can be queried and linked to analysis software, and how reports will be generated.

The structural design of the database involves taking the specifications from the needs assessment portion of the conceptual design and joining these needs to the correct "automated processors" (Yourdon 1989). This step involves designing the appropriate tasks and interfaces for the database (Yourdon 1989). For example, if the database must produce reports based on the calls for service involving incidents of residential burglary in each police district, the first step is to establish how the calls for services are handled. If, for example, the operator writes the information down then this hard copy report format is the first process that needs to be automated. This is done either by having the system automatically determine the address and/or by recording the call and entering the data into the computer at another time. The design of the database for the calls is important. For example, categorizing the data from the calls for service is helpful: the time (minute, hour, day, weekday, month, year) of the call, the time of the incident, the location of the caller, the location of the incident, and the type of incident should be recorded (coded for a category of incident, such as use of a numeric value for auto accidents, emergency services requests, or crime incidents).

A database dictionary contains the elements in the database, explains what is in the database and how the fields are linked (if at all) in the database. The dictionary will specify the format of the data, such as character, numeric, or date formats for example. The data dictionary creates "an organized listing of all the data elements that are pertinent to the system, with precise, rigorous definitions so that both user and systems analyst will have a common understanding of all inputs, outputs, components of storage, and intermediate calculations" (Yourdon 1989, 189). This helps to clarify the "lineage" of the database, that is, where the data came from.

The manner in which the data is entered is also part of the system design, such as the interface the user uses to input the data. There are several methods for structuring the data fields that can be used in designing the data entry process that assists in producing an efficient database. These methods include using numeric and text descriptions, input masks, validation rules, pick lists, look-up tables, and a selection of unique identifiers (USDOJ 1997b). In any case, an interface should be designed in such a manner that it allows for ease of data entry and data consistency.

The design of the data entry component must also consider whether or not the database is networked with several different users simultaneously entering data or if the users are independently entering data into a table that is appended to a localized database. Quality control measures must be introduced if there is more that one user who can access the database, to prevent data redundancy and corruption.

A decision should also be made to determine how the different data tables would link together. Normally, agencies assign a number or code that represents the crime incident or call for service, and this variable is used as a unique identifier that can help link the related records in various data tables together. It is important that all crime data records contain this unique ID (primary key) if there are different tables for the incident, victim, suspect, modus operandi, and/or recovery location. In addition, since the database is being designed as a spatial database, an address or location must be included in at least one table. Figure 4.1 shows an example of how some of the data tables in the database might be linked.

How the database is to be used in conjunction with analysis programs must also be determined. Some software packages allow for a direct import of the data tables into analysis packages and extensions. If the database is

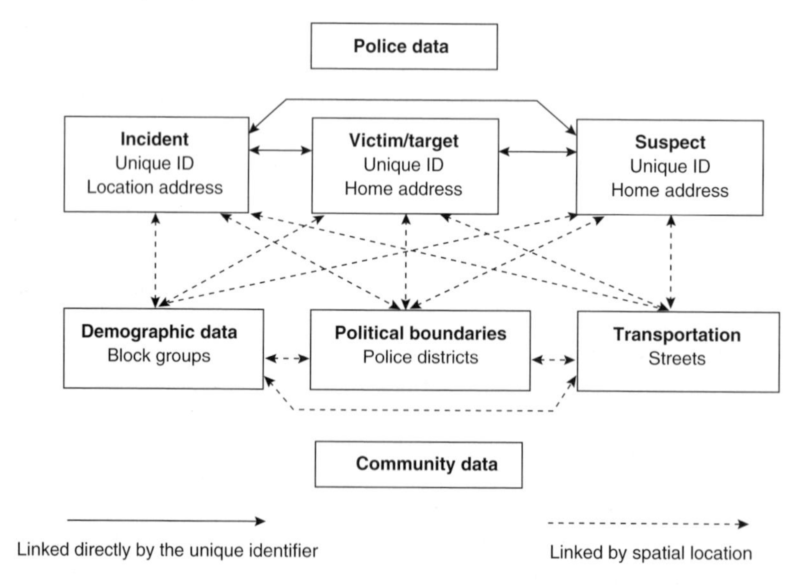

Figure 4.1 Linking data tables within a spatial database.

a large one, choice of an open database connectivity structure may be appropriate. Particularly, if there is an interface that allows the analysis software access to the database through a structured query language (SQL). This all depends on the characteristics of the software that is used for both the database and the spatial and statistical analysis.

Also needed is a method to interpret the voluminous information that is capable of being generated by the GIS (Block 1997). This can be done by designing a standardized method for generating reports. The standardized report is based on what is most vital for typical users of the system to receive on a routine basis.

Physical design

Finally, the physical design of the system can begin. An inventory of existing hardware, data storage and input devices, and networking capabilities should reveal what is needed. This stage establishes how the database is housed, accessed, updated, and backed up. Often the hardware already exists within the agency. Computers, networks, modems, printers, flatbed scanners and data storage and backup devices are commonplace among many agencies.

Hardware for a GIS database first includes a computer to house it. The storage capacity of computers has increased dramatically over the past years; external storage devices are becoming unnecessary for uses other than data back-up and for the transport of data. The number of computers needed is going to depend on the number of users, purpose of the GIS and the frequency with which the database is accessed.

Graphical input devices, such as a scanner, may be needed. In most instances, law enforcement agencies can collect a lot of the base data from other agencies. For example, the public infrastructure data and political boundaries may already exist within a city's engineering department, so the creation of new data for a base map would not be necessary. If such base data does not exist, an agency may not have the personnel resources for this form of data entry and would then consider outsourcing this work and in any case converting scanned data into a topological vector data structure may be beyond the technical abilities of many departments. However, items such as photographs of crime scenes, evidence, or reports maybe scanned into the computer and linked to a location.

The connectivity of the database to other users or networks should be assessed in this phase. It should have been determined in the logical design phase how the data was going to be accessed and shared by different users or agencies. It also should have been established whether or not this database would serve as a public resource, and could be accessed over the Internet. Establishing these functions would show the type of networking needed and whether or not to connect by modem, cable, or even wireless connections such as to mobile data terminals. This would also be related to the security of the database, which is discussed subsequently.

There are a couple of different output devices. A printer can be used for printing reports, summaries of crime in specific areas, and a host of other information from the database. There are also larger printers, called plotters, which can be used to print large maps. The function of the agency and the uses for the GIS should dictate whether or not an agency really needs a plotter.

Quality control

Quality control is a very important element in database development and design. If there is inaccurate data entered into the database, inaccurate results will be generated in analysis. Standardization of all data input is also necessary. If for example the database has been designed to use military time as the format, the data must be consistently entered in as such, or any temporal analysis will become suspect.

Many of the database software programs allow the creation of customized data entry screens (report templates). For instance, if the user entered data into the database for the time that the incident occurred, a prompt could be issued from the database to remind the user that time must be entered in as a specific format. Address error entry might also cause difficulties for later geo-coding. A data dictionary of acceptable addresses can be created that will help flag erroneous addresses prior to either geo-coding failure or mis-located geo-coding. The database could also be designed in a manner that requires systematic data entry. An example of this would be to require all addresses to have a street number (closest to the incident) and street suffix (i.e. Oak St. or Oak Ave.). If the incident occurred at an intersection, both streets might need to be included. A further method would be to have a method for identifying whether or not the incident occurred in the building (that would mean that the address would be more accurate) or outside, near the given address.

Methods also need to be developed for incidents that occur where there are no addresses or cross streets. In these cases, the agency could establish a methodology where the law enforcement agent reporting the crime could potentially note from the scene of the incident the mileage and direction from the scene of the incident to the next intersection, and these incidents could then be manually geo-coded into the geographic data layer. Alternatively, GPS could determine accurate absolute locations for such incidents in a less cumbersome fashion.

Security

There will be many people who will need access to the database for different reasons, and it will often be up to the systems manager of the database to decide who will have what level of access, if any at all. This type of authority involves deciding who can view, change, update, and delete the data as well as any combination of these permissions.

There are many different ways to secure a database. One method of security that is often overlooked is the physical security of the premises. The systems manager may want to consider a method that makes sure only those with access to the database can get into the computer room or other areas with terminals that can access the database. This type of security can be carried out by allowing only those employees with user authorization access to the predefined areas, or by properly identifying visitors. The next level of security is identification tags and passwords. Identification tags are a method to control access. For example, the Chicago Police Department uses a card key scanner peripheral device on every computer. Office identification cards carry a bar code that is read by the reader. Thus, the system controls access and records the duration of use and identity of every system user as well as which databases are accessed and from which workstation the access was made. Passwords go a step further in allowing only authorized users into certain areas within the system, such as the database. There is also user-level security which, presumably, regulates who has access to modify the data in the database.

The systems manager is cognizant of the security of the data. Data security can be achieved by encryption of data files in the database. This is necessary if there are parts of the database that should not be viewed by all users. This can be achieved with user security programs. Theft of the data can be prevented if the systems manager sets up the database in such a way that users cannot duplicate or replicate the data. In addition, methods for preventing the destruction or loss of data such as a schedule of back-ups should be instituted.

Officially verified data and unofficially verified data should have two different security levels. According to Richard Block, official data should have "gone through a review process, usually standardized in format, have standard codes and an identification number of some kind, and are often considered public information. Whereas unverified investigation data such as lists of suspects, contacts or citizen tips might require high security" (Block 1997, 50).

Maintenance

If the system has been designed well, it should be flexible enough to incorporate new functions as the needs of users mature. An aspect of system maintenance is also making sure that the users are properly trained in the use of the technology. Additionally, the systems manager should establish how often the database is being used, and obtain feedback from the users on how the database is functioning. This type of feedback from the users can show functions that are not operating properly or new functions that are needed. Other events, such as the addition of a new agency within a multi-agency network, may require modification of the database which will affect the operation of the database likewise adoption of new approaches such as the "geodatabase" model involved in the latest generation of object oriented GIS software may require modification of the database schema. Maintenance, changes, and upgrades to the database should always be documented in order to insure optimal performance.

References

Arvanitis, L. G., B. Ramachandran, D. P. Brackett, H. Abd-El Rasol, and X. Du (2000) Multiresource inventories incorporating GIS, GPS and database management systems: A conceptual model. *Computers and Electronics in Agriculture* 28: 89–100.

Block, R. C. (1997) The GeoArchive: An information foundation for community policing, 27–81. In: D. Weisburd and T. McEwen (eds), *Crime Mapping and Crime Prevention. Crime Prevention Studies*. Monsey, NY: Willow Tree Press.

Burka, J. C. (1999) Breaking down jurisdictional barriers: A technical approach to regional crime analysis. *Proceedings, 1999 Environmental Systems Research Institute International User Conference*. Available at http://www.esri.com/library/userconf/archive.html.

DeMers, M. M. (2000) *Fundamentals of Geographic Information Systems*. New York: John Wiley & Sons.

Gorr, W. L., A. M. Olligschlaeger, J. Szczypula, and Y. Thompson (1999). Forecasting Crime. Grant No. 98-IJ-CX-K005, National Institute of Justice, Office of Justice Programs, US Department of Justice.

Harries, K. (1999) *Mapping Crime: Principle and Practice*. Washington, DC: US Department of Justice.

Nulph D., J. Burka and A. Mudd (1997) *Technical Approach to Developing a Spatial Crime Analysis System with ArcView GIS*. Washington, DC: US Department of Justice and INDUS Corporation.

Openshaw, S. (1995) Developing automated and smart spatial pattern exploration tools for geographical information systems applications. *Statistician* 44(1): 3–16.

US Department of Justice (1997a). Montgomery County, Maryland Police Department Spatial Crime Analysis System Project Three Month Review.

US Department of Justice (1997b). Montgomery County Department of Police Crime Analysis Section Analyst Tactical Database Documentation.

Vasiliev, I. (1996) Design issues to be considered when mapping time. In: C. H. Wood and C. P. Keller (eds), *Cartographic Design: Theoretical and Practical Perspectives*. New York: John Wiley & Sons.

Yourdon, E. (1989). *Modern Structured Analysis*. Englewood Cliffs, New Jersey: Yourdon Press.

Focus Box

From crime to focused distribution: Crime management data handling, South Yorkshire Police, England

Gary Birchall

GIS is presently utilized within the South Yorkshire Police (SYPOL). SYPOL is a regional policy agency in the north of England. GIS is a relatively new introduction within the constantly developing suites of software and hardware that any modern police force has to maintain. GIS assists the SYPOL in combating crime, but is also used to produce the required statistical data to show how well or badly SYPOL is performing against central government indicators.

SER world and crime management system

The GIS utilized by SYPOL is SER World 1.0. It is used to assist with the interpretation and analysis of data obtained from two of SYPOL's main databases, the COMRAD system and the Crime Management System (CMS). The CMS is a database, which is owned and operated by SYPOL, to capture and store recorded crime data within the County boundary of South Yorkshire.

For the purposes of the CMS database, an individual crime event is broken down into many component parts, so that effective analysis and subsequent representation can take place. SYPOL has over 400 crime classifications, over 200 of which are crime recordable offences for the Home Office (the Home Office is the central government ministry responsible for national crime assessment) counting procedures. A crime report is then separated into individual "fields" for data input purposes; these include property, documents, suspects, etc. which are all sub-fields to the main crime header which records the core details of the crime, time and date reported, time and date committed, person reporting, etc. All the respective fields within a crime report have been designed to be fully searchable, which assists analysis considerably.

The standards of data already held within SYPOL's CMS caused concern when the purchase of a GIS was first considered. Data on CMS was not inputted using a gazetteer facility (a look-up table of actual addresses), which would have given the system uniformity in relation to addresses entered onto the system. All the entries were free text entries and as a result there were many anomalies, such as different spellings of addresses, streets, and areas. A key component that was invariably missing from the database was the post code, a crucial element when attempting to geo-code data using postal code subdivision as is frequently done in the United States using ZIP codes data, although typically by market researchers. However, unlike the US, unique postal codes with corresponding x, y coordinates have been created for over 30 million individual residences and businesses in the United Kingdom. In the US not only are zipcode zone fairly large covering numerous residents, but their boundaries do not match other common mapping units like cities, counties or census tracts. In spite of the existence of georeferenced postal addresses SYPOL did not have a robust data collection and processing system that would be able to support a GIS.

To resolve these issues, it was decided that the database would have to be "cleansed" and brought up to an acceptable standard, which would allow data to be geo-coded and geographically represented, when a GIS was finally implemented. Temporary staff were assigned the task of manually checking over half a million CMS records stretching back to 1996, checking the anomalies previously stated which were then eradicated. Such retrospective geo-coding efforts are a rarity in most US police agencies due to lack of personnel as well as the low probability that cases other than homicides will be followed up after a year or more has elapsed.

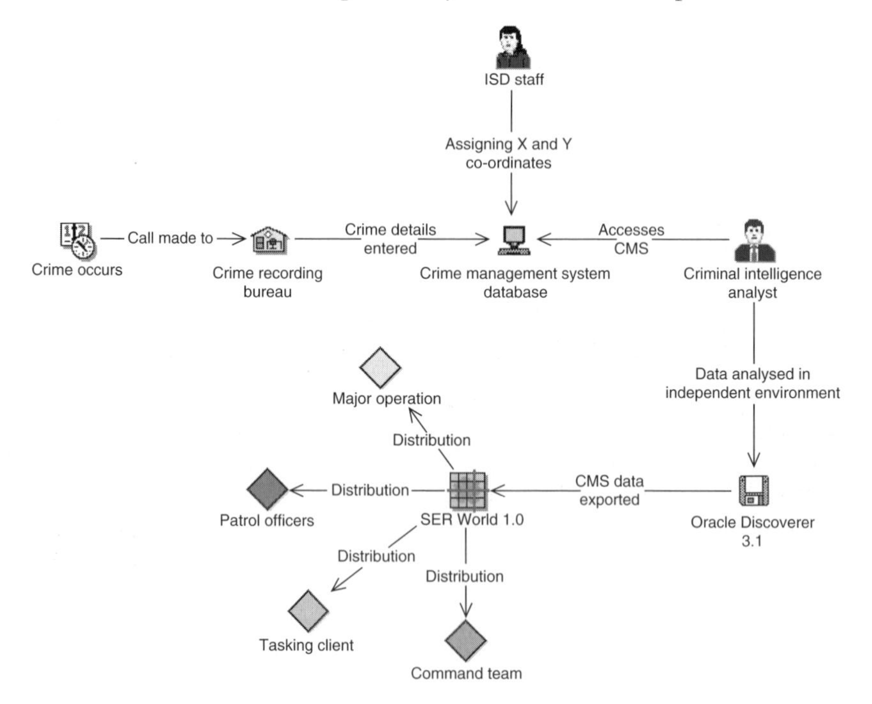

Figure FB4.1 From reported crime to focused distribution, CMS data handling and GIS processes within South Yorkshire Police.

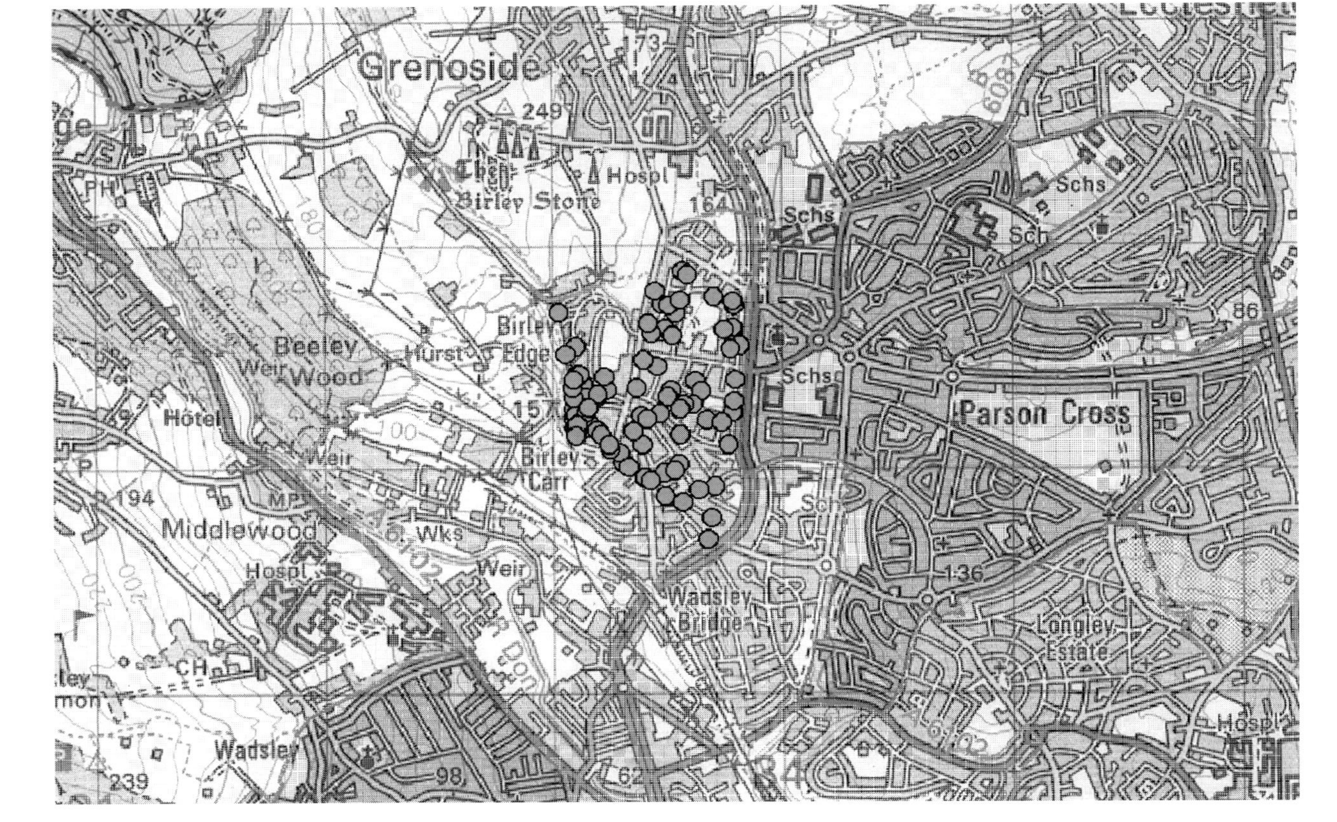

Figure FB4.2 Example of a detailed crime map created by the South Yorkshire Police's GIS. It uses the SER World GIS program and has a high level of detail including topographic features and streets represented by curb lines and not just centerlines. Geocoding is based principally on postal codes that define the geographic coordinates of virtually every home and business in England. (Reproduced by kind permission of Ordnance Survey © Crown Copyright NC/10162/02.)

Data handling process

Figure FB4.1 displays the data handling process now in place within SYPOL to geographically display incidences of crime within South Yorkshire. The assignment of geo-codes is at this time reliant on human intervention and inserts a data verification element into the process.

SYPOL has developed a more robust data handling system that is combined with adequate hardware capabilities. The Criminal Intelligence Analysts within SYPOL, now have a key role within the organization as they have ready access to important spatial information which can influence decision making at police districts, both at an operational and managerial level maps showing crime incident locations for individual residences based on postal code are displayed on the detailed Ordnance Survey GIS base maps that show numerous features including topography and building footprints as well as detailed streets portrayed by type and with varying widths (Figure FB4.2).

This process gives superior service to clients and the organization as a whole, with quick and efficient data handling and clarity of geographic representation.

5 Crime mapping and data sharing

Julie Wartell

Should law enforcement share crime maps and data? If so, with whom and how? Many people have strong opinions about this subject, although it is a relatively new issue. The controversy is caused by an oil and water mixing of existing public information and privacy laws, traditional law enforcement agency policies and philosophies, and a recent but growing demand for information combined with the technology to allow mass access and distribution.

Although a few law enforcement agencies started using GIS regularly in the late 1980s, it did not become widely used until the late 1990s. Most of the maps being generated were used for tactical, strategic, or administrative *internal* purposes. Prior to the World Wide Web, crime maps and data were being shared on a minimal basis, when a few agencies shared paper crime maps and spatial data with community groups, politicians, and researchers. In 1995, a small police department in California put the first crime maps on their Web site.[1] Others slowly followed, and it has only been in the last few years that it has become a controversial topic. Displaying crime maps on the Internet is the ultimate sharing (since they are available to the world), but there are other means for, and issues surrounding, the distribution of crime maps and data – across agencies or with researchers and other select groups.

Several events have already occurred in regards to discussion and resources for sharing crime maps and spatial data. Each of the last three National Institute of Justice Crime Mapping Research Center (CMRC) annual conferences has included sessions on this topic. The presentations and discussions have been informative and lively. In addition, in early 1999, "crimemap," the CMRC's listserv, had about twenty postings relating to the subject, and the CMRC held a two-day roundtable in July 1999 to discuss such things as providing data without violating privacy, and liability and security issues. Finally, the CMRC commissioned a monograph, *Sharing Crime Maps and Spatial Data: Meeting the Challenges*, that will be released in the summer 2001.[2] This chapter is not meant to be repetitive or redundant of the CMRC monograph but instead offers a concise overview of the issues and recommendations for law enforcement agencies implementing GIS.[3]

Why share crime maps and data?

There are a number of reasons why law enforcement agencies are sharing crime maps and data. These range from community and research partnerships and problem solving to a reduction in calls and requests for maps and statistics. The community has a right to know what is going on in its neighborhood, and this is an effective means of providing this information. If the community has the information, residents and businesses are more likely to want to work with the police to solve problems and maintain public safety. In *Sharing Crime Maps and Spatial Data*, the authors note the following advantages:

- Providing crime maps via the Internet or another convenient mechanism may actually reduce police workload; that is, fewer calls may be made to the crime analysis section for special requests if the maps are readily available.
- Many police departments have found that the more the community knows about crime and safety issues, the more willing it is to work with the police to solve those problems.
- Maps can assist in community policing and problem solving by showing where problems do and do not exist.
- Maps can increase public awareness about neighborhood problems.
- Maps facilitate partnerships with researchers and other agencies.
- By providing maps and data, a police department can be sure the data are presented accurately (as opposed to crime maps created by the media, private firms, or outside agencies).
- Providing maps and data to the public is a means to hold the police department accountable.

Although many feel the advantages of public release of crime maps outweigh the disadvantages, some potential negative issues still need to be mentioned. Critics of sharing crime maps and data with the general public via the Internet have a variety of concerns. Many say that providing crime information could re-institute the practice of redlining because the higher crime neighborhoods are often also the minority neighborhoods. Others are concerned that the information will be used for commercial purposes, such as alarm companies contacting victims of burglaries. Some believe that criminals might use the data to further their criminal behavior, either by being aware of where the police are likely focusing their attention (and avoiding those areas), or by knowing where "easy" victims are located (and targeting those areas). Lastly, there is a genuine concern for misinterpretation of data that leads to misidentification of the real problem. All of these are realistic concerns and should not be ignored. Instead, departments should discuss these issues when planning for sharing maps and data with the public and be able to respond accordingly.

What are the major issues?

If an agency is considering sharing crime maps and spatial data with the general public, researchers, or other agencies, there are a number of issues to be cognizant of. These include legalities, politics, societal effects, and security.

The legal issues can be very complex and are worthy of extra planning time and discussion with knowledgeable personnel. Privacy protections for victims vary across jurisdictions and need to be examined according to age, crime type, and who has access to the data (differences may exist between the public, researchers, and other law enforcement agencies). The privacy and safety of the criminal justice system personnel is also a slightly different issue.[4] Finally, there is the potential liability for the law enforcement agency or city that is providing the information. Some argue that liability can increase if something negative occurs based on the public having the information. Others counter that additional serious crimes might happen if potential victims are not made aware.

Disclaimers and confidentiality statements should be used to reduce the chance for liability. A number of police departments currently use disclaimers (see Sacramento as an example website, http://citymaps.sacto.org/GISAPPS2/cdisclaimer.htm). The US Census Bureau has an example of a confidentiality statement at http://factfinder.census.gov/html/confidential.html.

Politics often play a large part in a department's decision to begin providing maps and data to the public. Although the Chief might think it is a great idea to keep the community informed, local politicians may fear the negative repercussions (to tourism, business development, or residential influx) if a truer picture of crime in certain neighborhoods is revealed. Politicians may need to be convinced that the community has a right to know and that this can be done without violating victims' rights.

There are several potential societal implications of sharing crime maps and spatial data. On the positive side, a community could become more mobilized and involved in partnering with law enforcement and other entities to identify and solve public safety issues. On the other hand, critics of sharing information believe releasing crime data could have an adverse effect on property values and business stability in a neighborhood where crime is more concentrated. In addition, some are anxious that mass dissemination of crime information will have racial implications or provide more of an opportunity for repeat victimization. Although these have yet to be proven, agencies need to be aware of these concerns.

Security issues are always a concern with law enforcement information, but vary according to who will have access, and to what type of data is being shared. For publicly accessible web sites on the Internet, access is generally not controlled, unless there are specific groups that need to retrieve different information than the general public. In the latter case, passwords or encryption should be implemented. If the maps that are being made available are graphics (without a database behind them), data security is not an issue. If the web site has interactive mapping, and hackers can get to the data, departments should take precautions such as use of firewalls, duplicate databases, removal of victims' names or exact addresses, and security software. In sharing spatial data and maps with other law enforcement agencies and researchers, security procedures need to be put in place, but will depend on laws and department policies. Research should be done, and security experts should be consulted, during the planning process for sharing data.

Where should we begin?

Before going forward with providing crime maps and spatial data to the public, researchers, or other law enforcement agencies, current partnerships and web sites should be assessed. The reasons for an agency wanting to make this type of information available as well as specific objectives and limitations (e.g. laws, policies, budget and expertise) must be identified. Explore what other agencies are doing as it relates to agency objectives, and the "good" and "bad" features of each. Create a plan, based on thorough research, for implementation. The plan should include what is desired to be done, why, and how it will be targeted, marketed, and the site will be maintained for prospective users.

In regards to sharing spatial data across law enforcement agencies, the US Department of Justice's strategic approaches to community safety initiative (SACSI) project provides a good example (http://www.ojp.usdoj.gov/nij/pubs-sum/sacsi.htm). Five sites were funded to develop a multi-agency, GIS-based system for community safety problem solving. Other examples include the "Enforcer Geographic Information System" for Pinellas County, Florida law enforcement agencies (http://www.co.pinellas.fl.us/bcc/juscoord/enforcer.htm) (see Chapter 13) and the Regional Crime Analysis Geographic Information System developed by the Department of Justice's Criminal Division for thirteen agencies in the Baltimore, Maryland region.

The idea of sharing crime data and maps with researchers is growing rapidly. Many researchers have established working relationships with law enforcement agencies, and the element of trust already exists. Sometimes researchers are given direct access to law enforcement agency systems, and other times they receive exports of the spatial data for their own use. In the planning stages, both the researchers and the law enforcement agency need to discuss what data is necessary, means for accessing it, how it will be used, and where and how it will be maintained during the project, and whether destroyed or stored upon completion. One excellent example of a spatial data sharing arrangement was the Hartford Comprehensive Communities Partnership. This effort, between a police department, a researcher, and a community organization, entailed providing the community organization with raw calls for service, crime, and arrest data in order for the partners to map and analyze the data for themselves.[5]

As mentioned in the beginning of this chapter, the CMRC of the US Justice Department in Washington, DC, maintains a list of Internet sites that have crime maps and data available to the public. There is a wide range of information and means for display and dissemination, and all should be examined. Exemplary sites include those of San Diego County's ARJIS (www.arjis.org) which is an interactive site that covers all law enforcement agencies for the county. Cambridge, Massachusetts (http://www.ci.cambridge.ma.us/~CPD/crime/index.html) also puts crime patterns and series on the web. One of the newer sites was created by the Chicago Police Department (http://12.17.79.6/). It is very interactive and targets the citizens of Chicago to partner with the police in problem solving. Lastly, Milwaukee (http://www.gis.ci.mil.wi.us/Map_Milwaukee/map_milwaukee.htm) has a limited amount of crime information, but includes an immense amount of other data.

Whether your plans include sharing spatial data and maps with the public, other agencies, or researchers, there are state laws and federal policies that exist regarding privacy, confidentiality, and release of public records. In addition, with respect to research conducted under a federal grant there are further rules such as the protection of human subjects,[6] privacy certificates, and confidentiality requirements.

Prior to sharing spatial data and maps with the public, other agencies, or researchers, many decisions need to be made relating to the data itself as well as the map products. One of the first questions to be asked is what type of data will be shared. Also, will it be specific point or aggregate data, how accurate will it be and how current is it, and how should it be explained.

In regards to the type of data, most departments are making crime or call for service data available, although some also provide arrest and citizen complaint information. Within the previous categories, the agency must then determine which crime, call, and arrest types can legally be made public, and which will most benefit the various users (this will differ greatly between researchers, other law enforcement agencies, and the general public). Many departments avoid providing juvenile or sexual assault information to the public, as well as information that is part of an ongoing investigation. In addition to crime data, other types of data that might be shown on a map are: schools, parks, hospitals, public housing, alcohol licensees, and crime-prone businesses such as banks, convenience stores, malls, and hotels.[7] Other considerations include whether the maps have point data or shaded reporting areas. There are advantages and disadvantages to each. One means for providing specific point information without violating victim privacy is to identify locations by the generalized hundred-block address (e.g. 1234 Main Street would show up as 1200 Main Street).

The accuracy of data that is being made available to the public is extremely important. When the maps are produced, agencies need to consider that most users are not educated map readers and are not likely to know how to interpret complex maps (such as certain graduated symbol and chloropleth maps). The data needs to be defined (what is a burglary, what is the difference between a call for service and a crime), and interpretation guidelines can be very helpful. Geo-coding rates should be shown and explained. If only 75 percent of the crimes are being displayed on the map, there is a great deal of room for misinformation and misinterpretation; unfortunately, "hit rates" are rarely released.

Cartographic design issues are essential but are often neglected. These issues range from the type of map to disseminate (static or interactive), to symbology, to the legend. Static maps (unchangeable graphics) are easier and cheaper to produce but do not allow the users to decide what they want to see, whereas interactive mapping permits the user to make selections such as date range, crime type, and geographic area. Symbols can make or break the map. If the symbols are all squares and circles of different colors, a person who prints the map in black and white will not be able to differentiate the crime types. Also, the size of the symbols relative to features such as streets will affect how big or little the crime problem is for an area. This relates also to the scale. A scale helps users determine how concentrated the crime is in their neighborhood. The legend should explain all symbols on the map. Finally, clear visual references with labels, such as major roads, schools, and parks, are important for people to orient themselves, particularly in larger cities and those with gridiron street patterns.

Summary and resources

The crime mapping field is still developing, and agencies are continually shaping policy and practice when it comes to sharing crime maps and spatial data. This chapter is intended to bring the topic to the forefront of the planning process for agencies implementing GIS; it is not meant to be an all-inclusive how-to guide. The best recommendation is for agencies, and the personnel assigned to the task of implementing GIS, to spend time reading through the existing literature, evaluating other agencies' web sites and policies, and talking with as many stakeholders as possible. In other words, doing a careful assessment of who is currently sharing crime maps and data, how they are doing it, and who in your community can have an effect or will be effected by this effort.

There are a number of resources available for agencies considering sharing crime maps and spatial data. In addition to the ones listed below, a great deal of information is available in GIS literature and other (non-law enforcement) governmental publications:

- List of agencies with crime maps and data on the web (www.ojp.usdoj.gov/cmrc/weblinks/welcome.html)
- Federal Geo-Spatial Data Committee (FGDC) policies and standards (www.fgdc.gov)
- Publications

- *Sharing Crime Maps and Spatial Data: Meeting the Challenges* (Julie Wartell and J. Thomas McEwen, forthcoming from NIJ); draft at www.ilj.org
- Notes from NIJ Crime Mapping and Data Confidentiality Roundtable (CMRC, 1999)
- *Mapping Across Boundaries: Regional Crime Analysis* (Nancy La Vigne and Julie Wartell, forthcoming from Police Executive Research Forum)
- *Crime Mapping News*, quarterly publication of the Police Foundation.

Notes

1 Putting crime on the map. *Police*, June 2000.
2 A draft version is currently available at www.ilj.org.
3 Further "Resource" details are included at the end of the chapter.
4 This issue has been dealt with in the courts relating to property information being provided on the Internet through the Tax Assessor's Office. See Ryan Thornburg's article "GIS and the Privacy Puzzle," *Governing*, December 1999.
5 More detailed information about this project is highlighted in *Crime Mapping Case Studies: Successes in the Field, Vol. 1*, Police Executive Research Forum, 1998.
6 See NIJ's Regulations for the Protection of Human Subjects at www.ojp.usdoj.gov/nij/humansubjects/hs_02.html.
7 Most agencies have very limited non-crime data beyond street center lines and jurisdictional boundaries. These are examples of what could also be useful on a crime map.

6 GIS and crime mapping by Illinois police: If you've got it, flaunt it

J. Gayle Mericle and Kenneth Clontz

The evolution of using the spatial analysis capabilities of GIS for the mapping of crime would seem to be hitting high gear these days, with more and more police departments using GIS as a major weapon in their planning arsenal. The push to move from the traditional reactive law enforcement response to a proactive, potentially preemptive approach to fighting crime has prompted agencies to pinpoint the geographic location of incidents and act upon the information. This change is all very heartening and positive, but there it falters. A majority of departments studied reserve information gathered from GIS for *in-house* use, declining the opportunity to pass on what they have learned to the public they serve. Police department web pages abound with glitzy graphics, public relation homilies to persuade the inquiring citizen of the agency's user-friendly nature, and in one notable case, a toggle feature allowing the viewer to have a cartoon officer perform a dance to the theme of the "Cops" television program. But specific information regarding the actual location of illegal events in the community is hard to come by, at least for the Internet-using public.

GIS and the Internet: nationwide trends

What prompts this look at the Illinois police departments' Internet pages is an earlier, exploratory study (Clontz and Mericle 1999). Here, the authors began to look at law enforcement sites nationwide to determine how agencies were using GIS-based mapping on their sites, and if they were used, what type of displays were presented for public access (Table 6.1). A secondary research question examined whether different parts of the country varied in their use of the Internet and their type of GIS mapping display. Original findings showed that 97 percent of the United States police agencies with sites did not provide any GIS maps online. Of the few departments displaying any maps, the most common types shown were static in nature. The two regions of the country that lagged furthest behind in Internet use and GIS mapping for community inquiry were the Midwest and Southeast, respectively.

As Table 6.2 shows, Illinois in 1999 was noteworthy as having 96 agencies having sites of any kind. Of the sites that could be accessed, 94 pages

Table 6.1 Type of map

Name of department	*Name of state and region*	*Displays static maps*	*Displays dynamic maps*	*Displays animated maps*
Brea	CA – Far West	X		
Oxnard	CA – Far West	X		
Redondo Beach	CA – Far West	X		
Sacramento	CA – Far West	X		
Salinas	CA – Far West	X		
San Francisco	CA – Far West		X	
Santa Rosa	CA – Far West	X		
Torrance	CA – Far West	X		
Mesa	AZ – Southwest	X	X	X
Scottsdale	AZ – Southwest	X		
S. Tucson	AZ – Southwest		X	
Boulder	CO – Southwest	X	X	
Colorado Springs	CO – Southwest	X		
Douglas County	CO – Southwest	X		
Longmont	CO – Southwest	X		
Illinois State Police	IL – Midwest		X	
Overland Park	IL – Midwest	X		
La Grange PD	GA – Southeast	X		

Table 6.2 Number of law enforcement web sites identified by search engines in Illinois

Year of study	*Number of web sites identified*	*Number of departments without GIS maps*	*Number of departments with GIS maps*
1999	96	94	2
2001	128	125	3

did not show any GIS maps. Only two departments used and displayed crime maps for their public. With the passing of three years it was wondered if, at least in Illinois, these numbers had changed.

Crime mapping/GIS and the Internet: Illinois law enforcement agencies

This section examines how law enforcement agencies in Illinois use and/or fail to furnish crime information plotted by GIS mapping on their Internet sites to their citizens. Of the approximately 1,200 agencies in that state, only the Illinois State Police and local police agencies were selected. County agencies were also initially examined, but of the 102 in Illinois, only 22 maintained sites. After examining those, Sheriff's departments had to be excluded from the study as not one of the county agencies with an Internet site presented or mentioned using GIS mapping. In all, 106 Illinois police departments' sites were examined for use and type of presentation of GIS mapping. The results were not encouraging.

Dynamic maps are the most prevalent type of map displayed by Illinois law enforcement agencies as of 2001, a clear change from the State's past displays. Table 6.3 shows the type of map and the number displayed at each site.

Departments seem to use a variety of ways to identify the location of their maps on their sites. Table 6.4 shows the various labels that departments use. The most popular seem to be "Crime Maps" or "Crime Statistics."

Many departments have embraced both GIS and the concepts of community and problem-oriented policing. Recognition that the existing quasi-military structure and philosophies prevalent in policing no longer serve the needs or expectations of modern communities led to this ostensibly whole-hearted partnership between citizens and the law enforcement agencies. So, it would seem reasonable that these community-oriented police organizations would invite public access to crime prevention tips and be willing to freely

Table 6.3 Type of GIS maps

Department	*1999 maps*		*2001 maps*	
	Static	*Dynamic*	*Static*	*Dynamic*
Chicago P.D.				X
Evanston P.D.			X	
Illinois State Police		X		X
Orland Park P.D.	X			

Table 6.4 How Illinois departments list GIS maps on their web sites

Department	*URL*	*Menu title on departmental web site*
Chicago	12.17.79.61	Citizen icon – can view data for the previous 14 days by address, beat, intersection, or school
Evanston	www.evanstonpolice.com	Report Document Viewer – user cannot change the information presented
Illinois State Police	www.state.il.us/isp/ishpagen.html	Interactive maps – crashes and crime

exchange vital data with their civilian "partners" in community-oriented policing.

What is clear from the data is a glaring contradiction of stated purposes. Few other stakeholders understand the commitment and requirements necessary for valid partnerships to flourish as well as do police officers. They know first-hand the value and need for the honest, complete exchange of accurate information and trust that must be established for a successful, even lifesaving relationship. Sampling of various Illinois police agencies' sites indicated that when it comes to applying these same principles of communication and partnership to interacting with the communities they rely on as important co-workers in establishing and maintaining public peace and safety, most departments fall mute. Vital information on crimes – type and location of incidents – suffers the fate of matter and energy venturing too near a black hole in space. Everything enters, but nothing escapes. The law enforcement community needs to be alerted to the apparent message they are sending of "do as I say, not as I do" when requesting that citizens provide detailed information to the police.

Sharing data: what's the problem?

One thing is clear from the data. A majority of Illinois police departments that are on-line to their public are still not freely sharing their crime mapping data. It is just as clear that these departments are using GIS for several in-house purposes, because they state they are doing so to illustrate the agencies' commitment to using technology to advance law enforcement effectiveness. There are a number of possible reasons why departments might not be freely displaying the data they are mapping.

One factor is probably simple oversight. Police agencies may not have considered the value of providing such information on their sites. Since crime is reported in other media for public consumption, it may never have occurred to those responsible for GIS crime mapping to take the next logical step and put up the charts in electronic format. What this amounts to, however, is failing to expend only a minor additional effort in a process that uses significant manpower. By showing the resulting crime maps to the public, the departments will be getting more for their money. Graphic presentation of crime information on maps gives the citizens a more comprehensive way to understand the incidents in their areas or zones than does the piecemeal, daily facts published in the newspaper or announced by the local radio station.

Another reason lies buried in the old, but still functioning philosophies inherited from American's policing quasi-military foundations. There was a time when departments serving communities only provided information to the public they served on a "need to know" basis or limited release of crime data to the newsperson in the "police beat." That time has passed. Few departments consciously adhere to that rigid, paternalistic style of policing, but may still function under that "eyes only" mind-set regarding the release of crime maps. This philosophical oversight still results in less than efficient use of time and resources expended on GIS mapping. In the case of GIS data, if the information is important to the agencies, then it is also important that the public be able to gain access to that data.

A third reason for not presenting GIS crime mapping to the public is due to technological limitations specifically inadequate bandwidth and slow access times. Lack of adequate bandwidth to translate the large graphic files that are notoriously slow to download can stop or discourage agencies from including such maps with their departmental information on their websites. Also, some departments may hesitate to incorporate the images for fear of alienating interested parties who quickly tire of waiting for the graphics to load.

Lastly, and most unfortunately, the specter of angering those with political power may deter some departments from making available information in the form of maps reflecting with inescapable clarity the disparate crime problems of some districts or zones. For example, real-estate people may not find this information on certain troubled neighborhoods advantageous for their businesses. Such people tend to be respected and powerful within their communities, often having the necessary clout with city councilpersons and others in positions of authority to make their complaints felt. Similarly "community leaders" in minority or disadvantaged areas may feel these maps pour salt on a wound rather than expose a festering sore to healing air. While there are several other external problems impacting on the departments' decisions, consider that police departments themselves may fear pressure by neighborhood groups dissatisfied with the number of criminal acts in their area, and blame the law enforcement agency for not meeting their needs. Add in concern over the issue of citizens' rights to privacy and/or confidentiality, and increasing national security related concerns in the wake of terrorist attacks on the US and incidents of organized hacking and it becomes understandable that some agencies prove reluctant to open this particular can of worms with public access of GIS crime-mapping data. "No access, no problems" seems to be their stance.

Conclusion

None of these issues raised is trivial. Each represents a major obstacle to sharing GIS crime mapping over the Internet with members of the community. Resolving the factors creating a barrier between what the law enforcement agencies know and what the public is permitted to learn is essential to fully involving populations of each jurisdiction as partners with their police. An uninformed partner, whether by accident or design, is a partner who cannot meet his or her responsibilities completely. And a partner who realizes he or she is excluded from receiving information that others (police personnel) routinely are given is a person who feels trust has been betrayed.

Providing data on incidents that have already occurred and showing a general location of such incidents do not cripple any police department or violate citizen privacy. As private security firms have learned, after years of being notoriously reticent in sharing accurate information about losses to the businesses under their protection, keeping such data in-house is equivalent to denying the existence of a problem. They have recently discovered that being honest and open about the incidents and their rates of occurrence will allow police to begin to realistically deal with the issues involved and their underlying causes (Gips 1999). Police departments can bolster their community policing efforts by being as open with their citizens as they are with their officers about where and how often crime happens. This openness encourages neighborhoods to understand what law enforcement has known for years. When it comes to criminal activity, the citizens should not see crime in their area as a case of "the police have a problem here." Making the results of GIS crime mapping accessible to the public enhances the perception that "we" – everyone in the community – have both a problem and a responsibility where crime is concerned.

References

Clontz, K. and J. G. Mericle (1999) Community policing: Put your mapping where your mouth is. Proceedings, 1999 Environmental Systems Research Institute International User Conference. Available at http://www.esri.com/library/userconf/archive.html.

Gips, M. (1999) High-tech manufacturers assess theft stats. *Security Management*, 11.

Focus Box

Comparing actual hot spots v. media hot spots: Houston, Texas, homicides 1986–94

Derek J. Paulsen

In reporting on crime, the media play an important role in shaping the public's perception of crimes, criminals, and the role of government in crime control. This focus box compares the locations of hot spots of actual homicides with incidents receiving either celebrated or local news coverage in Houston, Texas for 1986, 1988, 1990, 1992, and 1994.

Data collection

Official police records were obtained from the Houston Police Department covering the 4,980 incidents of criminal homicide between 1986 and 1994. These data contained the following variables: victim and offender name, incident location (street address), motive, victim/offender relationship, weapon used, and victim and offender age, gender, and race. These police records were then compared with a newspaper article database consisting of every homicide article published in the *Houston Chronicle* between 1986 and 1994. Importantly, this database was received directly from the *Houston Chronicle* in an electronic format, indicating that it is a potentially more reliable and accurate account of all homicide articles than a manual search of newspapers because it is the official list as known by the *Houston Chronicle* organization. Every homicide incident for the nine-year period was checked with the *Houston Chronicle* database to determine several important factors concerning news coverage. The following specific questions were addressed: Was there an article covering the homicide? If so, how long was it in both column inches and word length? How many articles were there about the incident? In what section and page of the newspaper was the article placed? Was there a picture with the article? In addition to these questions, other variable were examined, including the mention of race, incident location, motive, victim–offender relationship, weapon, and what stage of the criminal justice process the article covered. Simultaneously, all articles dealing with intimate homicides were separately analyzed by placement into a Word file. This qualitative file was then further divided into two separate files, one containing all articles covered within the first fifteen pages of the newspaper and the second containing all other intimate homicide articles. Finally, using the incident location provided by the official police data, these homicide incidents were geo-coded for use in spatial analyses within the study.

Methods

Hot spot analysis determines statistically significant concentrations of point patterns, much like a multivariate cluster analysis. Specifically, nearest neighbor hierarchical clustering (Nnh) determines groups of points that are spatially closer than would be expected to occur by chance alone. In conducting an Nnh hot spot analysis using Crimestat, a program developed by the Illinois Criminal Justice Information Authority, the user is required to specify two important criteria prior to beginning the analysis. First, the minimum number of points required for the generation of a hot spot is selected. In this analysis, the minimum number of points was set at five incidents for all four data subsets. Secondly, the user must select the probability level for defining the threshold distance between the points in the hot spot. This selection determines the probability that the hot spot could be due to chance alone. In this analysis the probability level was set at the 0.05 level.

Results

Two general observations can be made about the spatial distributions of the different hot spots over the nine-year study period. First, all of the hot spots for the different levels of newspaper coverage are spatially consistent with actual homicide hot spots. Regardless of the type of coverage or number of hot spots created, every single newspaper coverage hot spot overlaps or is contained within at least one actual homicide hot spot. However, it is important to note that not every actual homicide hot spot is spatially consistent with a newspaper coverage hot spot. This spatial inconsistency varies from a low of six actual hot spots with no corresponding newspaper coverage hot spot in 1986, to a high of twenty-five actual hot spots with no corresponding newspaper coverage hot spot in 1990. Overall, 54 percent of all actual homicide hot spots between the years 1986 and 1994 had no corresponding newspaper coverage hot spots. These results seem to indicate that while newspaper coverage hot spots are spatially consistent with actual homicide hot spots, they are very selective in which hot spots they correspond to. Importantly, it appears that those actual homicide hot spots that are closer to the city center are more likely to correspond to newspaper coverage hot spots than those on the periphery of the city.

The other important result from the hot spot analysis concerns the different location of the hot spots for celebrated coverage, local coverage and the crime column. While the difference in the total number of hot spots for each level of newspaper coverage can largely be attributed to the number of articles in each different category, the distribution of these hot spots spatially is not so easily explained. The three different categories of newspaper coverage hot spots seem to have very different spatial patterns. The trend across the study time period seems to indicate that the more celebrated the newspaper coverage the more focused the hot spots near the city center, with less celebrated newspaper coverage being spread out more on the periphery of the city. Celebrated article hot spots, although limited to only three total hot spots in the years 1986, 1988, and 1990 (Figure FB6.1), appear to be more focused near the city center. Similarly, local covered hot spots are focused near the city center in the years 1986, 1988, and 1992,

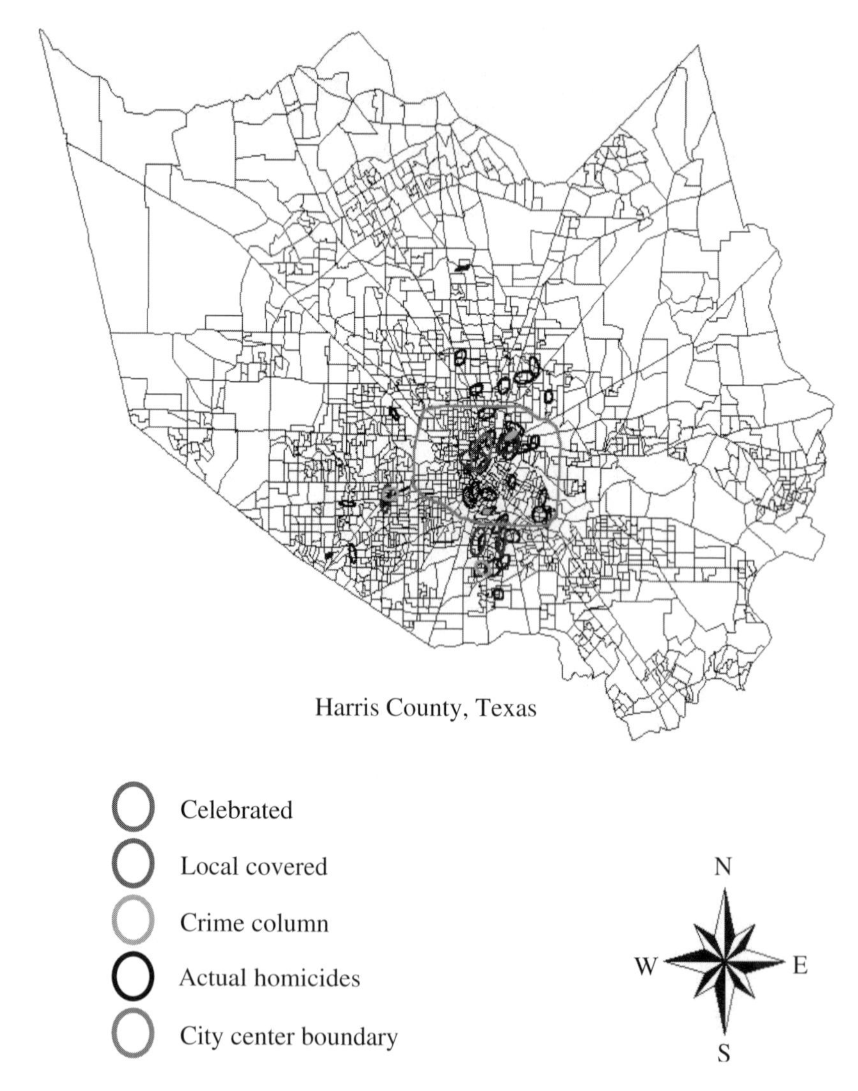

Figure FB6.1 Hot spot analysis for 1990.

but more spatially dispersed in the years 1990 and 1994 (Figure FB6.2). In contrast, the hot spot locations for articles in the crime column seem to correspond to more actual hot spots that are on the periphery of the city than either of the other two levels of newspaper coverage. Overall, this seems to indicate that celebrated newspaper coverage seems to focus on homicides near the city center, while less celebrated newspaper coverage (crime column) seems to focus more on homicides that occur on the periphery of the city.

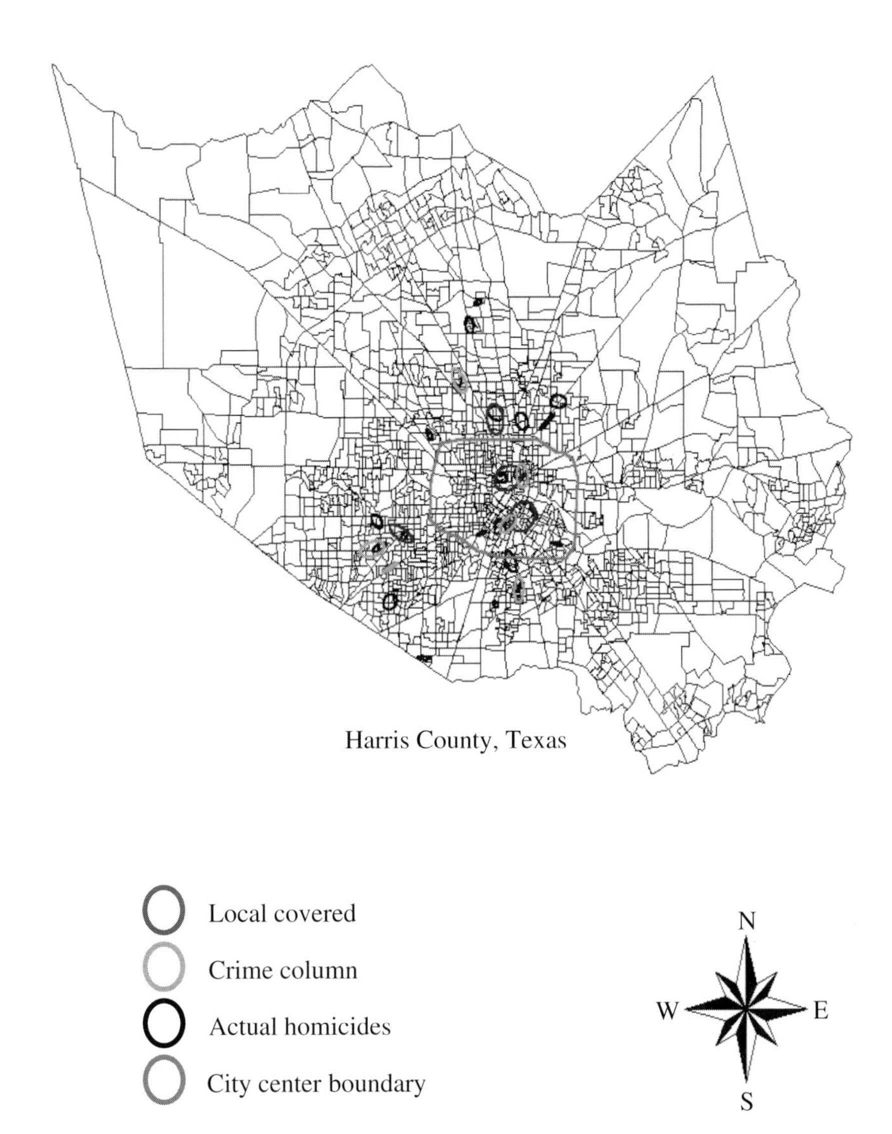

Figure FB6.2 Hot spot analysis for 1994.

Conclusion

Although newspaper coverage hot spots are spatially consistent with actual hot spot locations, they ignore the majority of actual hot spots, specifically those on the periphery of the city. Moreover, celebrated hot spots are more likely to correspond to actual hot spots near the city center than less celebrated hot spots that tend to correspond to actual hot spots on the periphery of the city.

7 Future directions in crime mapping

Andreas Olligschlaeger

There is little doubt that crime mapping has come a long way over the past decade. Ever since the days of simple pin mapping, law enforcement practitioners vendors and academics alike have sought ways in which to continuously improve upon the use of GIS in law enforcement. The establishment of the Crime Mapping Research Center by the National Institute of Justice about five years ago not only helped to legitimize the field of crime mapping within law enforcement and academia, but through its annual conferences and research grants also provided a means for practitioners, academics and vendors to communicate and share research and ideas. The result has been an almost unprecedented level of cooperation among all organizations interested in the field of crime mapping in the United States.

Compared to other, more established fields however, crime mapping is only in its infancy. For the most part, GIS is still primarily used as a standalone tool for digital pin mapping rather than for more sophisticated spatial and temporal analysis. This is not because other tools and methods for crime mapping do not exist, but for the most part it is due to the fact that many recent advances in crime mapping have yet to be incorporated into commercially available software. In addition, many crime analysts continue to undergo a daily struggle just to obtain and geo-code the data required for mapping, with little or no access to data warehousing facilities or even direct access to records management (RMS) or computer-aided dispatch (CAD) systems. However, these problems can all be attributed to "growing pains" and are not unlike those experienced in other emerging technologies. In fact, considering that crime mapping is a relatively small field, it is all the more surprising that it has come so far in such a short period of time. This is surely a testament to all the dedicated professionals that have worked hard over the past years to make crime mapping what it is today.

There is no doubt that crime mapping will continue to grow over the next decade. The question is: where will the next decade lead us? While recent crime mapping conferences have offered a glimpse of what lies ahead, the number of possibilities is boundless. Consequently, exploring all of the possibilities would be beyond the scope of this chapter. For that reason I will concentrate on two approaches that in my opinion show great promise for crime mapping.

Both approaches involve the integration of crime mapping with other tools and methods. The first, forecasting, takes crime mapping a step beyond the current practice of using GIS for pin mapping and as a tool for analysis of past crime trends. The second approach is geo-coding of free-form text via entity extraction. This method looks at ways in which a crime analyst can obtain data from non-traditional sources and in non-traditional formats.

Forecasting crime

For the crime analyst, it is important not only to be able to identify areas of high crime activity, but also to be able to anticipate where (and when) a crime is going to occur. Typically, the short term forecast horizon for a crime analyst is on the order of one month. The problem, though, is that short term forecasting is far more difficult than long term forecasting, especially when it comes to intra-urban forecasts for relatively small geographical areas (Gorr *et al.* 2000). In general, the smaller the geographical unit and the shorter the forecast time period, the more difficult it is to produce accurate forecasts. The primary reason for this is that there is simply not enough data per time period and geographic unit to produce statistically significant results; forecast estimates become too sensitive to noise in the data. Nevertheless, research has shown that, given enough data and the right model, forecasting can produce surprisingly accurate results.

While forecasting has been around for a long time, it is very new to the field of criminology as a whole and to crime mapping in particular. The pioneering efforts to incorporate forecasting with crime mapping were in part the result of a series of five parallel grants awarded by the Crime Mapping Research Center to researchers in the United States. Two of the five projects explored the use of forecasting methods for routine use by crime analysts. Brown and Dalton (1998) collected data on special events such as high school football games, end of school year and concerts as forecast variables. Gorr and Olligschlaeger (1998) aggregated monthly counts (by area) of selected lesser crimes (such as assaults and burglaries) obtained from police offense reports and 911 calls for service. The aggregates were then used as leading indicator variables to forecast one month ahead values of Part 1 crimes. One other of the five projects used long term forecasting techniques more appropriate for annual policy-level decision-making rather than short term manpower allocation or crime prevention purposes (Kelly and Field 1998).

Of the various types of forecasting methodologies to date, leading indicator models have emerged as the most promising for routine use by crime analysts. Basically, leading indicator models assume that certain factors are precursors to events, such as a robbery or an aggravated assault. Leading indicator models do not seek to forecast any one particular event, but rather to identify those areas (such as census tracts, grid cells, or patrol sectors) that will experience unusually large increases (or decreases) in criminal activity of a certain type during the next time period. This is in fact very similar to the underlying hypothesis of the "Broken Windows" theory (Kelling and Coles 1996; Sampson and Raudenbush 1997) which postulates that crimes of public disorder and "nuisance" crimes, as well as many more types of serious crimes are manifestations of a weakened capacity of a community to maintain order in public areas. Since crimes of public disorder are easily measurable, they can be used as leading indicators (i.e. precursor variables) for more serious crimes.

For example, let's assume that our goal is to identify emerging drug markets before they become fully blown hot spots. One way to measure drug activity in an area is to identify and map the number of calls for service for drug dealing since they are indicative of public perception of the area (Olligschlaeger 1997a). Note that drug arrests, on the other hand, are more of an indicator of police activity. In addition, many drug arrests occur in areas other than where the drug dealing took place. Mapping calls for service is fine for looking at past drug activity, but just using past levels of calls does not help us to identify previously unknown areas of drug activity. On the other hand, it has been shown that crimes associated with drug dealing, such as robberies, assaults, and burglaries tend to increase before an area becomes an established drug market. In addition, physical characteristics of a neighborhood (such as abandoned buildings, graffiti, broken and boarded windows) can also contribute to an area becoming a target for drug dealers (Olligschlaeger 1997b). Therefore, we want to collect information on those crimes that are associated with, and leading indicators of, drug calls for service.

Let's assume further that our forecast horizon is one month, that is, we want to produce one month ahead forecasts for each area. A simple leading indicator regression model that could produce these forecasts would be:

$$Yf_{i(t+1)} = \alpha + \beta_1 x_{1i(t)} + \beta_2 x_{2i(t)} + \cdots + \beta_n x_{ni(t)}$$

where t is the current time period, $Yf_{i(t+1)}$ is the forecast for the next time period for area (i), α is the regression constant, $x_{1i(t)}$ through $x_{ni(t)}$ are the leading indicator variables for area (i) and time period t, and β_1 through β_n are the estimated regression coefficients.

Models such as the above have been quite successful in forecasting crimes, although in order for them to work well it is very important to have large amounts of accurate data. Indeed, preliminary results have shown that the larger the dataset, the better the results. For this reason, most research to date has concentrated on large US cities where substantial amounts of data have been available. Also, larger cities are less likely to experience major shifts in crime rates as a result of factors outside the scope of the regression model such as local economic activity.

A successful GIS-based forecasting model could have great potential for crime mapping. Not only would it be possible to identify those areas that are expected to have large increases or decreases in crime on a map, but use could also be made of the regression coefficients to identify the exact locations that are potentially responsible for an increase in crime. For example, for the sake of argument, let's assume that the regression coefficients for two leading indicator variables – the number of abandoned buildings in an area, as well as the number of calls for service for aggravated assaults – are especially large and statistically significant. This implies that these two leading indicators play a major role in forecasts of large increases in the number of calls for service for street level drug dealing. Given this information, as well as point data on abandoned buildings and calls for service for aggravated assaults, a crime analyst could produce a detailed map of the area in question. In addition, the analyst might also want to overlay the residences of known drug offenders in that area, or the location of drug arrests. This map, along with copies of the relevant police reports, would be a very valuable source of information to investigators and assist them in proactive policing. Regression-based crime forecasting cannot be expected to yield accurate prediction in the face of major shift in criminal behavior such as the arrival of "crack" cocaine on the scene since the past is not always a good indicator of structured future trends in every situation.

Geo-coding via named entity extraction

A second highly promising application for crime mapping is geo-coding of free-form text using named entity extraction. Traditionally, geo-coding is a very formal affair. Precise information such as street number, street direction, street name, and street type is required for accurate geo-coding. In turn, this information is usually kept in database tables that are then address matched. Successful matches are inserted into a point layer (or coverage), which can be displayed on a map. The address information that is geo-coded is either entered manually or, as is most often the case, extracted along with other information from databases such as record management or 911 systems. In either case, the address data has to be highly structured, that is, follow a certain fixed format. However, many pieces of information exist that contain references to places that are not in a structured format. Examples include open source information such as newspaper articles as well as narratives from police reports or supplemental reports. In many systems only one or two addresses can be attached to, say, a police report or an offense report. But the narrative could contain many references to other places, for example, names of other gang members associated with a suspect. Being able to query a GIS for such information or map would be useful not only to crime analysts, but also to investigators.

Until recently it has been virtually impossible to automatically extract location information from free-form text and then geo-code it. Consider the following paragraph:

> The suspect, Bill Smith, was first observed exiting *Bill's Bar* in *South Breezewood* at the intersection of *Hilliard Ave. and Dunn St.*; *Bill's Bar*

is a known hangout of the Outlaw Biker Gang. Officers then followed the suspect to his residence at *34 Loudon Court.*

The paragraph contains four references to places: Bill's Bar (a place), South Breezewood (a town), Hilliard Ave. and Dunn St. (an intersection) and 34 Loudon Court (a residence). In addition, the paragraph contains a reference to a person (Bill Smith) and an organization (Outlaw Biker Gang).

Geo-coding the above paragraph poses a number of challenges. First, we must correctly identify those words that describe a place. In addition, we must distinguish them from other entities, such as people and organizations. For example, the word "Bill" occurs three times, once as part of a name and twice as part of a location. The technique used to correctly identify people, places, and organizations in free-form text (in addition to other information such as dates, time, and numbers) is known as entity extraction (Kubala *et al.* 1998). Originally used to extract information from speech-recognition, the technique was later expanded to include other forms of text such as newspaper articles and video (Christel *et al.* 2000). Entity extraction can be highly accurate (greater than 90 percent correct extraction rate). Based on speech-recognition techniques such as Hidden Markov Models (HMMs), entity extraction techniques use supervised training and Bayesian probabilities to associate word pairs and triplets with a particular type of entity. The following sentence illustrates how this works:

Michael Jordan flew to Jordan.

There are two instances of "Jordan" in the above sentence. A well trained entity extractor is able to properly distinguish between the fact that "Michael Jordan" refers to a person, and "flew to Jordan" refers to a place. The HMM parses the sentence using a Viterbi Search algorithm (which is basically a one forward pass probability optimizing algorithm) and for each word decides what type of word it is (person, place, organization, or other). The key is in the word pair and triplet probabilities. For example, the first instance of "Jordan" is preceded by the word "Michael." The probability that the current word is a person, given that the previous word is "Michael" is much higher than that of a place. Conversely, the probability that the current word is a place given that the previous word is "to" is much greater than that of a person or other type of entity. Hence, the entity extractor should correctly extract the person "Michael Jordan" and the place "Jordan" from the sentence.

Once all entities have been extracted, the next challenge is to geo-code the places that have been identified. Unlike traditional address matching, place information extracted from free-form text is much more difficult to geo-code. Perhaps, the most difficult part is that not all locations are addresses. For example, in the paragraph above we have one regular street address, one alias for a street address (Bill's Bar), an intersection, and a town.

Geo-coding must therefore first determine what type of location has been extracted. Next, if the coverage against which we want to geo-code contains more than one instance of an extracted location, we must somehow decide which is the correct location in the coverage to choose as a match.

One way to accomplish this is to look for other clues in the text. For example, let's assume that in our study area there are two instances of intersections between Hilliard Ave. and Dunn St. Re-parsing the sentence from which the intersection was extracted we find that there are two other locations mentioned. In this case the key is South Breezewood. Overlaying the South Breezewood municipal polygon with our address coverage we find that there is indeed an intersection between Hilliard Ave. and Dunn St. located there. Thus, in all likelihood we have the correct match.

A geo-coding process using entity extraction techniques very similar to that described above was used in automatically geo-coding video of CNN news broadcasts as part of the Informedia project at Carnegie Mellon University (Olligschlaeger 2001). The geo-coding success rate, including resolving ambiguous place names, was around 90 percent for approximately 2,000 hours of news video matched against 78,000 places worldwide (Olligschlaeger 1999). Geo-coding the video allows users to query for news segments from a map as well as automatically create maps from news stories, similar to the way in which CNN shows background maps (Christel *et al.* 2000; Hauptmann and Olligschlaeger 1999).

Conclusion

The techniques discussed in this paper represent just two of many promising new technologies emerging for use in crime mapping. However, the key to continued progress in the field of crime mapping is the successful integration of mapping with other methods. Both techniques require a substantial amount of preprocessing, which in turn requires considerable computing power and complex database schemas. In addition, they require access to data warehousing facilities, not only for preprocessing but also for subsequent map-based display and query of results. Nevertheless, initial research indicates that these techniques can work very well and strengthen the analytical capabilities of the field of crime mapping.

References

Brown, D. E. and J. Dalton (1998) *Spatial-Temporal Criminal Incident Prediction: A New Model.* Washington, DC: Predictive Modeling Cluster, Crime Mapping Research Center, National Institute of Justice.

Christel, M. G., A. M. Olligschlaeger, and H. Chang (2000) Interactive maps for a digital video library. *IEEE Multimedia Computing and Systems*, 60–67.

Gorr, W. L., A. M. Olligschlaeger, and Y. Thompson (2000) Assessment of crime forecasting accuracy for deployment of police. Carnegie Mellon University,

Pittsburgh, PA: Working Paper, H. John Heinz School of Public Policy and Management (forthcoming in International Journal of Forecasting, Special Issue on Crime Forecasting).

Gorr, W. L. and A. M. Olligschlaeger (1998) *Crime Hot Spot Forecasting: Modeling and Comparative Evaluation.* Washington, DC: Predictive Modeling Cluster, Crime Mapping Research Center, National Institute of Justice.

Hauptmann, A. and A. M. Olligschlaeger (1999) Using location information from speech recognition of television news broadcasts. Paper presented at the ESCA ETRW workshop: "Accessing Information in Spoken Audio," Cambridge University, England.

Kelly, W. and S. Field (1998) *A GIS Analysis of the Relationship Between Public Order and More Serious Crime.* Washington, DC: Predictive Modeling Cluster, Crime Mapping Research Center, National Institute of Justice.

Kelling, G. L. and C. M. Coles (1996) *Fixing Broken Windows: Restoring Order and Reducing Crime in our Communities.* New York, NY: Free Press.

Kubala, F. *et al.* (1998) Named entity extraction from speech. *Proc. DARPA Workshop on Broadcast News Understanding Systems*, Distr. By Morgan Kaufmann, copyright by DARPA, 287–292.

Olligschlaeger, A. M. (1997a) Crime mapping in the next century: An artificial neural network based early warning system. In: David Weisburd and Tom McEwen (eds) *Computerized Crime Mapping, Crime Prevention Series.* Camden, NJ: Rutgers University Press.

Olligschlaeger, A. M. (1997b) *Spatial analysis of crime using GIS-based data: Weighted spatial adaptive filtering and chaotic cellular forecasting with applications to street level drug markets.* Carnegie Mellon University, Pittsburgh, PA: Doctoral Dissertation, H. John Heinz III School of Public Policy & Management.

Olligschlaeger, A. M. (1999) Geocoding video using entity extraction: The Informedia Project. Paper presented at the ESRI Annual User Conference, San Diego, CA.

Olligschlaeger, A. M. (2001) Criminal intelligence databases and applications. In: M. B. Peterson and R. Wright (eds) *Intelligence 2000: Revising the Basic Elements* International Association of Law Enforcement Analysts (IALEIA) and Law Enforcement Intelligence Unit (LEIU).

Sampson, R. J. and S. W. Raudenbush (1997) Neighborhoods and violent crime: A multilevel study of collective efficacy. *Science* 277: 918–924.

Part II
Case studies

The second part of this book deals with specific cases of successful application of GIS in law enforcement agencies. In soliciting contributions for this book, the editors sought to obtain material from various sized departments, in a range of states and several countries. Also departments with responsibilities at the municipal, metropolitan (major city and surrounding county), county, regional, state, and national levels were contacted. Numerous law enforcement agencies responded and some of the material they provided were absolutely outstanding; all was appreciated.

Cases illustrating use of GIS by Municipal police departments in Lincoln (Nebraska), Knoxville (Tennessee), Spokane (Washington), Phoenix (Arizona), and Boston (Massachusetts) greatly extend material in focus boxes on use of GIS in municipal departments in Waco, Texas, and Overland Park, Kansas. Material on GIS use by the Pinellas County Sheriff's Department and local law enforcement agencies within that county and by the Delaware State Police, the National Guard Bureau Counter-drug office, and the South Yorkshire Police provides a perspective on use at other levels of government. These cases illustrate the application of GIS to a wide range of criminal justice, crime analysis, and police administrative and deployment issues. Examples of use of GIS to combat such crimes as sexual assault, robbery, burglary, and community disorder are provided. Non-traditional applications of GIS to event planning, traffic enforcement, crime scene, investigation automobile accident reconstruction, substation location, and redrawing jurisdictional boundaries are also provided. GIS is being used as a tool by crime analysts to solve specific series of robberies, by researchers to analyze travel to crime and to study homicide-related issues. More broadly, GIS is being applied by departments featured in these case studies as an integral part of community-oriented policing to highlight crimes affecting communities and focus resources such as outreach, to organize neighborhood watch programs and to help generate awareness for the need to provide higher levels of funding for after-school activities in crime blighted communities. So, taken together, these cases illustrate the broad range of potential applications of GIS and related technologies such as GPS and remote sensing. They clearly show that these technologies are playing an increasingly vital role as we move into the new century.

8 Lincoln Police Department – specific examples of GIS successes

Tom Casady

The following cases provide specific examples of the success of GIS in reducing crime in Lincoln, Nebraska, and managing resources within the Lincoln Police Department (LPD). There are just under 300 sworn offices in the LPD serving a population of 213,000. These cases reflect the diverse range of functions within a municipal police department that can be made more efficient through application of GIS. It is perhaps instructive to note that by targeting issues besides major crimes such as traffic safety, drunk driving and public drunkenness, the LPD has been able to achieve the distinction in 2001 of making Lincoln the major metropolitan area in the US with the lowest rate of drunken driving related accidents; this is in spite of being the location of the University of Nebraska campus.

(Editor)

Reducing residential burglaries in southeast Lincoln

During the period from 21 April through 5 May 1999, Officer Matt Tangen organized and carried out a problem-oriented policing project, after he became aware of a trend of residential burglaries occurring in four police reporting districts. The modus operandi in these burglaries was distinctive. The suspects were entering open garages and stealing high-value items such as bicycles and golf clubs. In some cases, the suspects entered vehicles parked inside garages, and took items such as cell phones and compact disk players. Officer Tangen and his fellow officers on Southeast, the Team B Beat, patrolled the affected districts looking for open garage doors and delivered an informational flyer to homeowners with open doors, warning them of the trend and suggesting precautions. During the project, officers located eighty-eight open doors in the target area. An analysis of the eight weeks following the project, compared to the same period in 1998, showed that the number of residential burglaries was reduced by two thirds, from thirty-one in 1998 to ten in 1999. GIS technology helped identify the geographic trend in these reporting districts, and also aided in the assessment of the results achieved.

Combating high-risk drinking parties near University of Nebraska – Lincoln

This project was part of a larger strategy of NU Directions, a campus-community coalition to reduce high-risk drinking by college students, one of ten similar coalitions funded by the Robert Wood Johnson Foundation. As part of the community strategy, the LPD sought to change college students' perception of the risk of being arrested or cited for alcohol-related violations such as: minor in possession of alcohol. Greater enforcement of alcohol prohibitions in campus housing, and more stringent enforcement of regulations in taverns had caused an increase in high-risk drinking at off-campus private parties. Since one or two officers could do little when confronted with a group of 50–200 party-goers, the perception had developed that the police would do little when responding to party complaints, and that the risk of arrest was small. Using GIS to identify neighborhoods close to campus where substantial numbers of complaints were received, this project focused a group of seven officers working on overtime to respond to such complaints during the fall semester of the 1998–99 school year (Figure 8.1). A rather small total of ninety-one citations were written, but the department publicized the project extensively, both before and after implementation. Using GIS, LPD analyzed the number of complaints

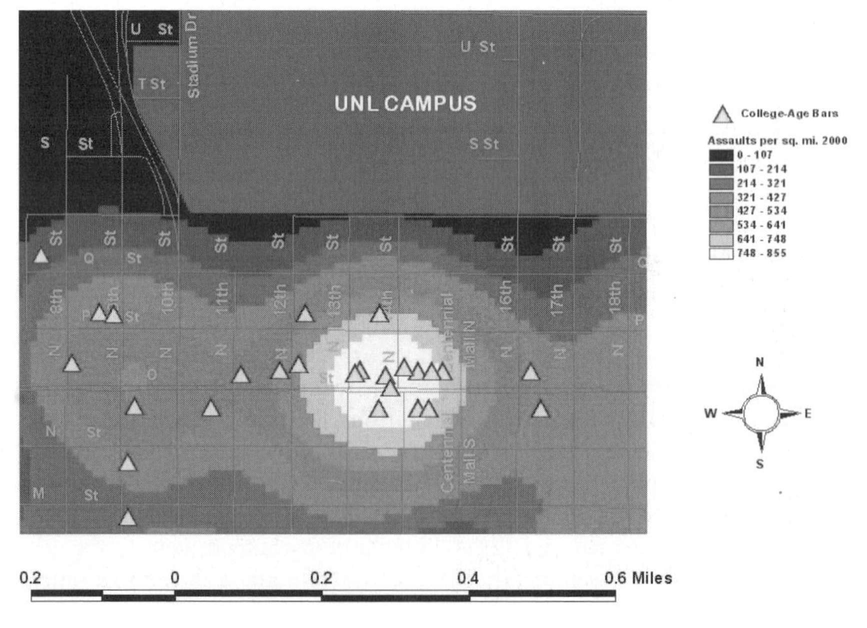

Figure 8.1 Combating high-risk drinking parties near UNL.

received from the public concerning wild parties, and found a 27 percent reduction in complaints from neighborhoods located within one mile of the campus, as compared to the same time period in 1997.

Creating a new community police team

During 1997, the LPD was considering modifying its major operational districts, referred to as Team Areas at LPD, in order to adapt to substantial population growth in the 20 years since the boundaries were last modified. An initial proposal to move one boundary line by six blocks was analyzed by the LPD Planning and Research staff – a project that required nearly a week. Rejecting that proposal due to the disparities in workload and staffing that would result, the planned redistricting was shelved for several months. In the spring of 1998, however, the department began examining options once again. This time, the availability of GIS software dramatically eased the process. At an April retreat, commanders worked interactively with census tract maps and tables including both geographic units and measure of workload, such as part I crimes and calls for service within each census tract. Maps and tables were projected on a screen, and dozens of options were examined to determine the impact of various "what if" scenarios on workload and staffing. Ultimately, the staff recommended the creation of a new team area (Figure 8.2). GIS maps were subsequently used to

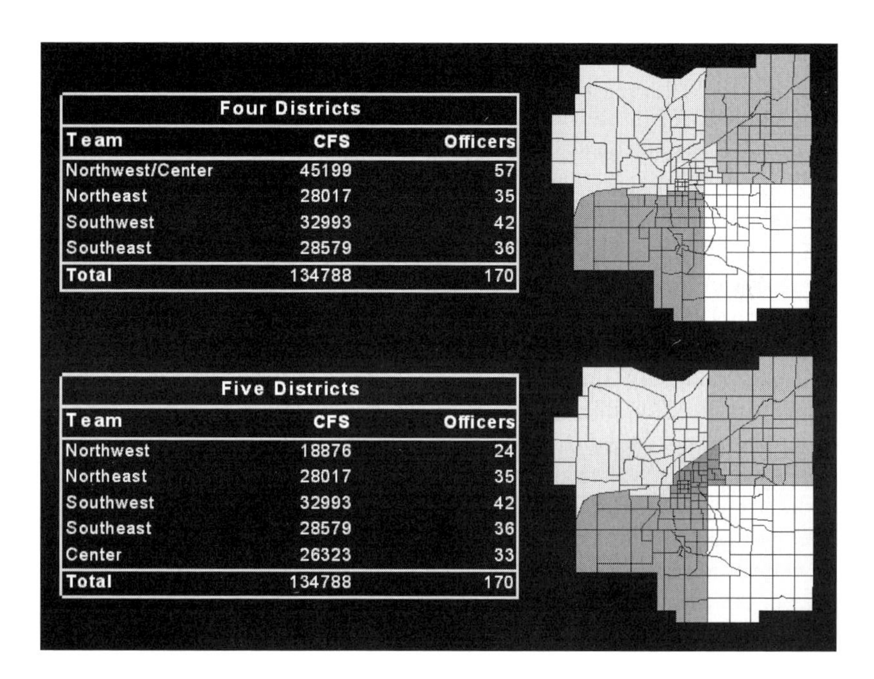

Four Districts		
Team	CFS	Officers
Northwest/Center	45199	57
Northeast	28017	35
Southwest	32993	42
Southeast	28579	36
Total	134788	170

Five Districts		
Team	CFS	Officers
Northwest	18876	24
Northeast	28017	35
Southwest	32993	42
Southeast	28579	36
Center	26323	33
Total	134788	170

Figure 8.2 Creating a new community police team.

seek the input of both elected officials and the general public. Additional modifications resulted, but the new team area was implemented on 7 January 1999. The comprehensive and objective analysis of every suggested alternative solidified a consensus within the department and community that was essential to the success of the redesigned team areas.

Solving the sexual assault of a child

On 21 March 1999, the LPD investigated the sexual assault of an 8-year-old boy in a field near the Far South Neighborhood in southwest Lincoln. This child had been walking home from a playmate's house across an abandoned railroad right-of-way and field when an assailant accosted him. The assault caused a significant community furor, and particularly frightened the residents of the adjacent housing subdivision and apartment complex. Examining other sex offenses in the same vicinity using GIS, investigators learned that several indecent exposures had occurred in the same general area over the course of the preceding 15 months. A suspect who shared a similar physical description and a common modus operandi committed these five offenses. Based on the pattern discerned by further analysis of the case files, officers initiated a surveillance of Wilderness Park, an area around which the previous cases were centered. One week after the sexual assault, officers located and arrested a suspect in the park, a 39-year-old convicted sex offender who was taken into custody wearing no pants, and subsequently convicted of the assault. GIS technology not only assisted in identifying the spatial relationship between each of the incidents and their proximity to Wilderness Park, but digital orthophotography displayed in GIS plots in a format that assisted investigators in organizing and conducting the surveillance project. The case also served as the impetus behind a departmental project which quickly released neighborhood crime information to the public via the Internet. GIS maps provide navigation to this data, and interactive maps can be created and viewed by the public at this site.

Coordinating crime scene investigations

GIS has become an important tool in LPD's crime scene investigation process. In virtually all major crimes, the department's crime analysts prepare projects containing streets, parks, parcels, aerial photos, and other features of the vicinity of the crime scene. Using an HP750 plotter, 36″ × 36″ layouts are produced, and these are displayed on white boards in the criminal investigations conference rooms. These accurate and detailed layouts take the place of blackboard drawings that investigators used for decades to depict the area of crimes while planning crime scene work and follow-up investigation. Similarly, large layouts of fatal traffic accident scenes are used to supplement scale diagrams and photographs. Prosecutors typically use these layouts as well during trial preparation and presentation.

Reducing police incidents at an apartment complex

On 17 September 2000, Officer Charles White responded to yet another complaint of a loud drinking party at the Claremont Park Apartments, on North 9th Street a few blocks from the University of Nebraska campus. On this occasion, the mood turned ugly, and some of the 200 plus members of the crowd pelted the officer with bottles and debris. Although no one was hurt, it marked just the latest in a series of disturbances at the complex, some of which had required that every single officer available in the city be dispatched to the area to quell the problems. Using GIS with geo-coded incident report and dispatch records, the department determined that over fifty disturbances had occurred at Claremont Park in the preceding year. On the following Tuesday, the police chief and area commander met with the complex manager, and made a number of suggestions for methods she could use to reduce the disturbances. This meeting was followed by an on-going dialog, and by further police contact with the complex's management company, in Lawrence, Kansas. As a result of a number of these interventions, the number of police incidents fell dramatically and immediately. GIS helped us identify the true extent of the police incidents at Claremont Park, and to assess the change in demand for police services over time (Figure 8.3).

Stopping a series of automobile break-ins

During the first week of December 2000, LPD crime analysts noted a concentration of thefts in the Meadowlane neighborhood in northeast Lincoln. The cluster consisted of thirty-seven automobile break-ins in one month within an area about ten square blocks (Figure 8.4). Alerted to the existence of this crime series, the officers assigned to the late shift on the Northeast Team began working special patrols and conducting surveillance in the neighborhood. Late on 8 December, just a few days after the trend had been identified, Officer Kelly Williamson stopped a suspicious vehicle leaving the neighborhood. The car had no license plates, and was just seen a few blocks from a reported prowler that other officers were responding to. She arrested the occupants, three young men who had been involved in the thefts. A considerable amount of stolen property was recovered from the car, and from the residence of one of the defendants, following the service of a search warrant. In the following six weeks, only one theft occurred in the neighborhood.

Planning special events

Every fall, on several Sundays, the University of Nebraska's Memorial Stadium becomes the third largest city in the State of Nebraska as the Corn Huskers football team attracts 78,000 fans to the latest in a string of sellouts that goes back to Lyndon Johnson's Presidency. Planning the traffic and crowd control details, involving assignment of over 50 officers to traffic direction

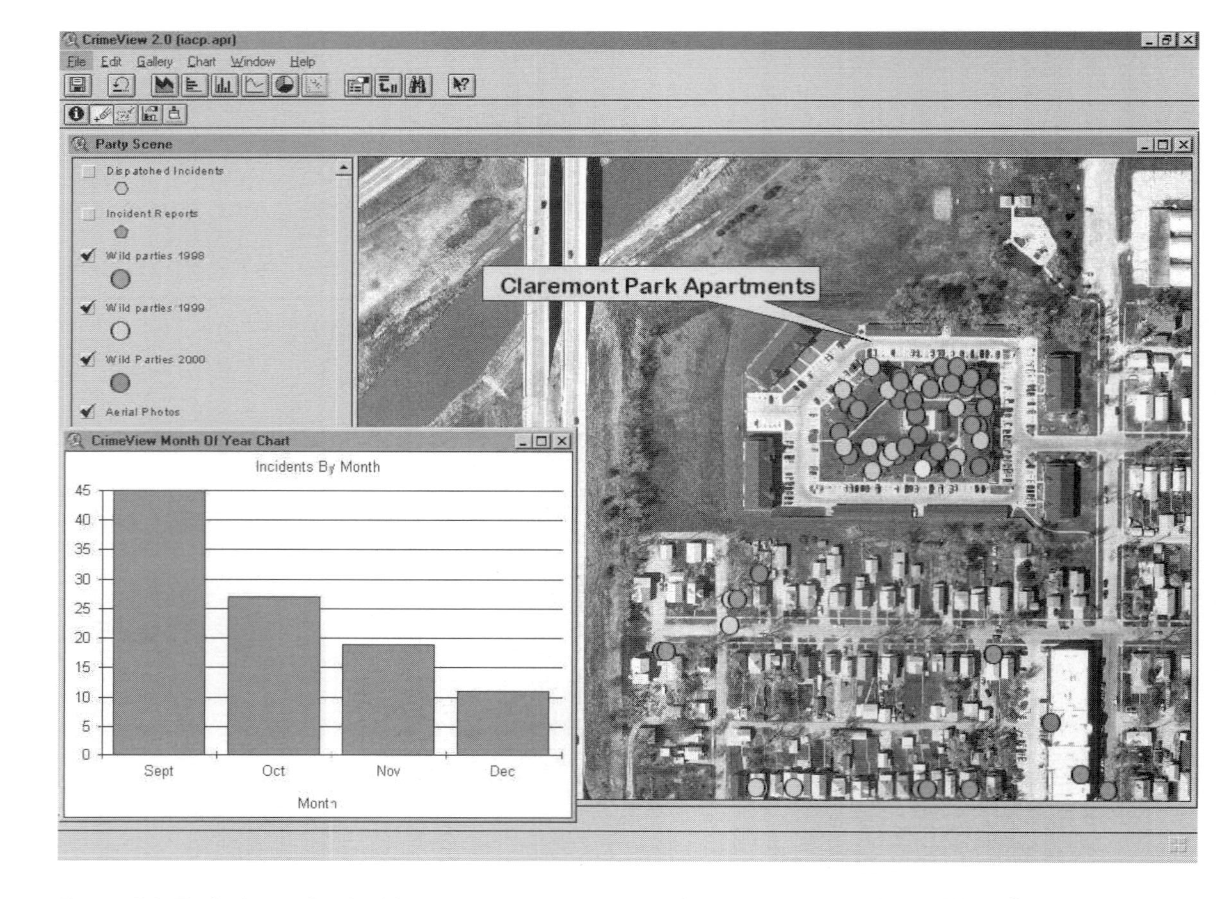

Figure 8.3 Reducing police incidents at an apartment complex. An example of Crime View® software in use.

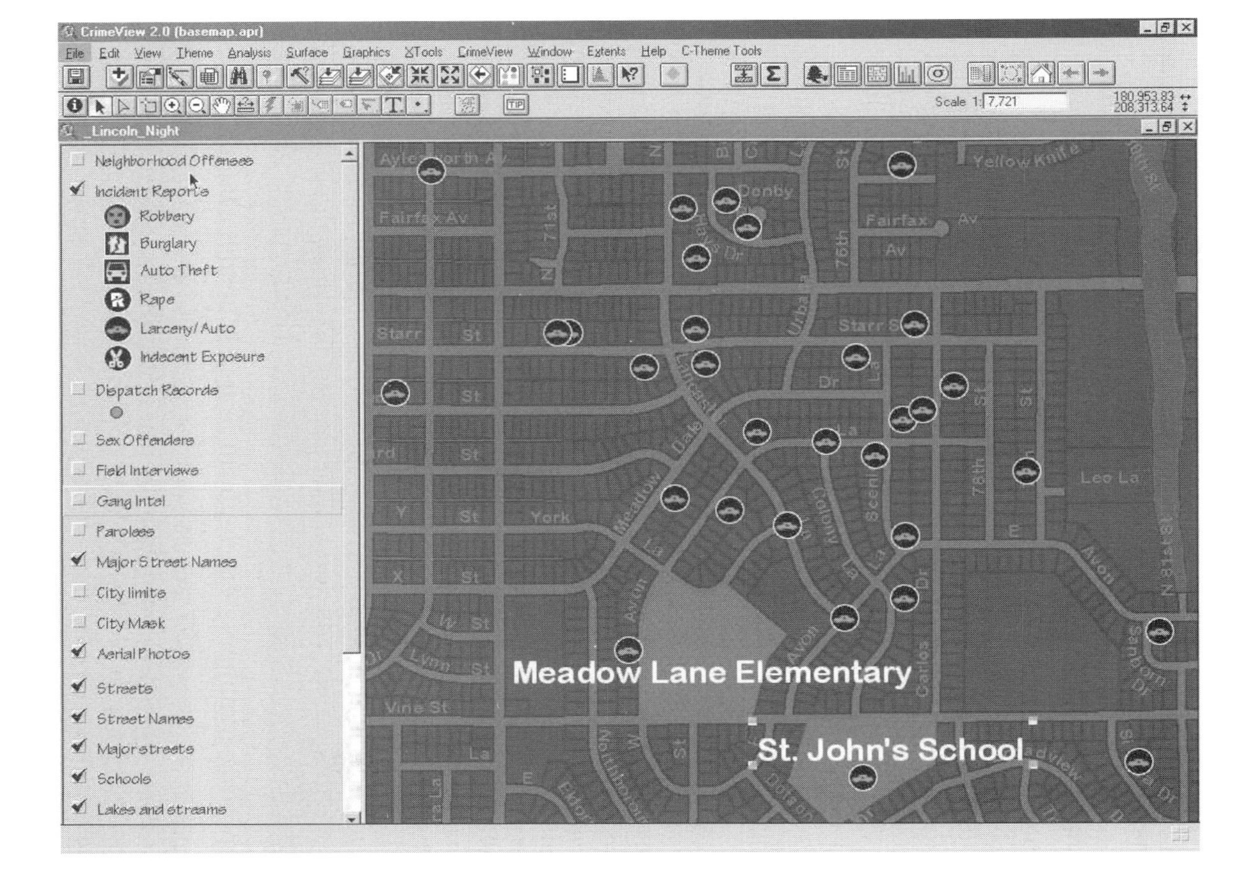

Figure 8.4 Stopping a series of automobile break-ins. CrimeView® copywrite the Omega Group, Inc.

points, is a monumental task. In recent years, GIS has assisted this process immensely. LPD managers use GIS-based analysis incorporating aerial orthophotos, building footprints, pavement and street markings to pinpoint assignments, design traffic patterns, and even specify the location of individual traffic cones. Similarly, GIS analysis and products are used for planning at other major special events, including the city's annual Independence Day celebration, Star City Holiday Parade, and the Lincoln Marathon.

Reducing thefts of self-service gasoline

In the fall of 2000, police managers noted a rather alarming increase in the number of larcenies. Examining the data more closely revealed that one type of theft – larcenies of gasoline from self-service pumps – was responsible for the vast majority of the increase, up nearly 400 offenses, an increase of over 80 percent from the same period in 1999. Using GIS, an analysis was conducted to identify the geographic extent of the offenses. A graduated symbol legend based on the count of incidents at individual gas stations revealed that a group of two-dozen stations and convenience stores were particularly hard-hit. One, the Gas-n-Shop convenience store at 951 West O Street, had 148 thefts during 2000, more than doubling the next closest location. Using these data, officers organized problem-oriented policing project beginning in November. This project was aimed primarily at a preventative approach – working with businesses to implement procedures that minimize the risk of drive-offs. These strategies include increasing staffing during peak theft hours, removing posters and sale bills from windows to improve visibility, using intercoms to welcome customers at the pump, installing video surveillance cameras at pump islands, encouraging pay-before-pumping policies, and other methods. November and December thefts plummeted, but the long-term impact of the strategy is not yet evident. A front-page newspaper article about the problem featured LPD's graduated symbol map, and caught the attention of a Nebraska State senator, who has introduced a bill in the 2001 session that increases penalties for this offense and revokes the drivers' license of those convicted (Figure 8.5).

Supporting community development in fragile neighborhoods

In the 2000 session of the Nebraska Legislature, the General Affairs Committee was considering legislation that would allow the City of Lincoln to use monies from Nebraska's Metropolitan Infrastructure Reinvestment Fund to help underwrite the construction of a new full-service recreation center in a fragile, high-crime neighborhood in south Lincoln. The planned center would replace the City's F Street Recreation Center, located in an antiquated and undersized century-old fire barn, and managed by Carla Decker, a very committed Parks Department employee. Testimony before the Committee by citizens and public officials concerned the vital need for the

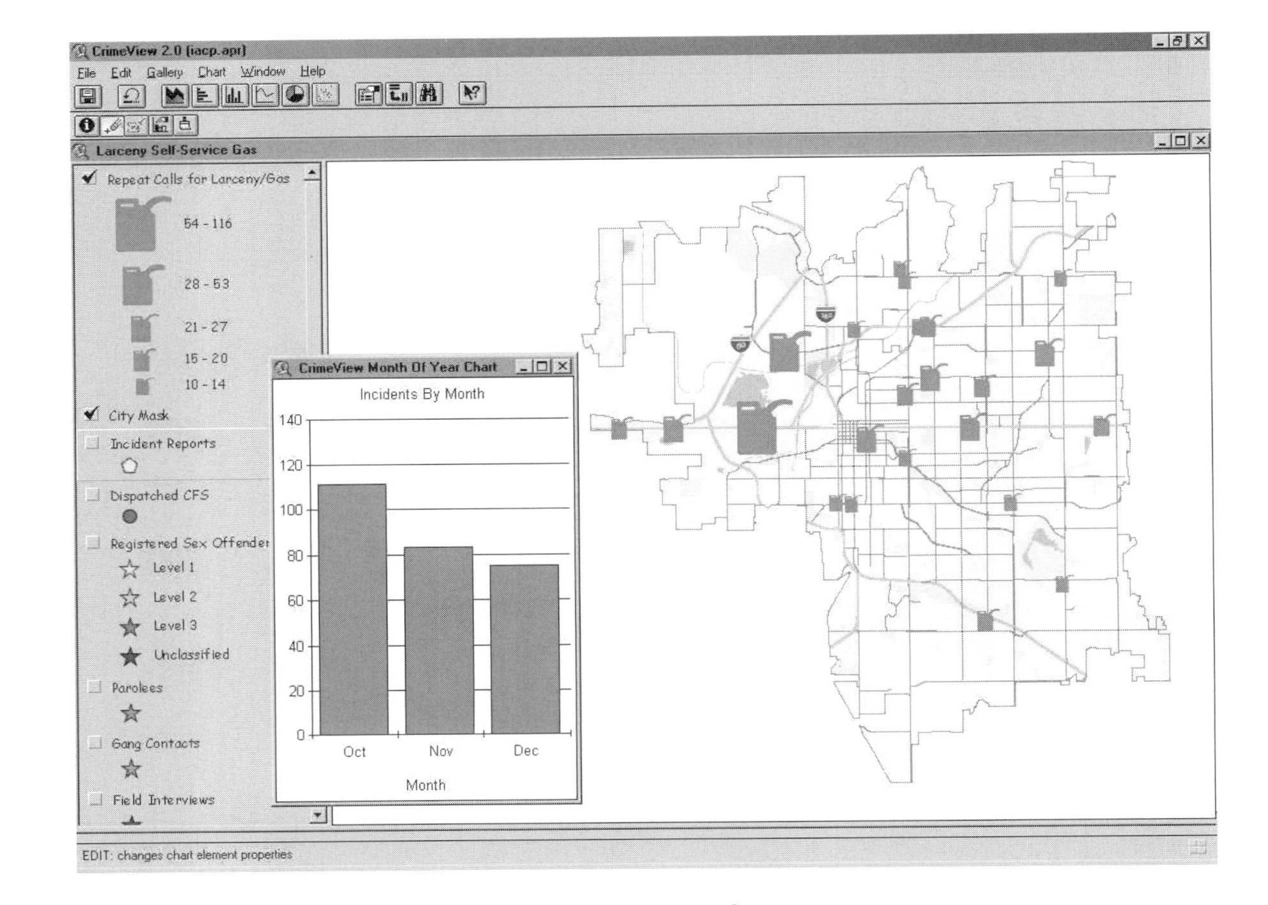

Figure 8.5 Reducing thefts of self-service gasoline. CrimeView®

Center to continue and expand its services to the surrounding neighborhood. Police Chief Tom Casady's testimony focused on the relationship of crime to the environment, and the need to concentrate on prevention resources in the areas of greatest need. The Center's services – after school care, safe recreational opportunities for all ages, inter-generational mentoring, senior services, literacy classes, and many others – are exactly the kind of investment in social capital needed to help stabilize this transitional neighborhood. A GIS map was prepared to demonstrate the concentration of crime in the neighborhood. The map used a simple design with a density grid of violent crime designed to draw the senators' eye and pique their interest (Figure 8.6). The needed legislation was enacted by the Legislature, and today construction of

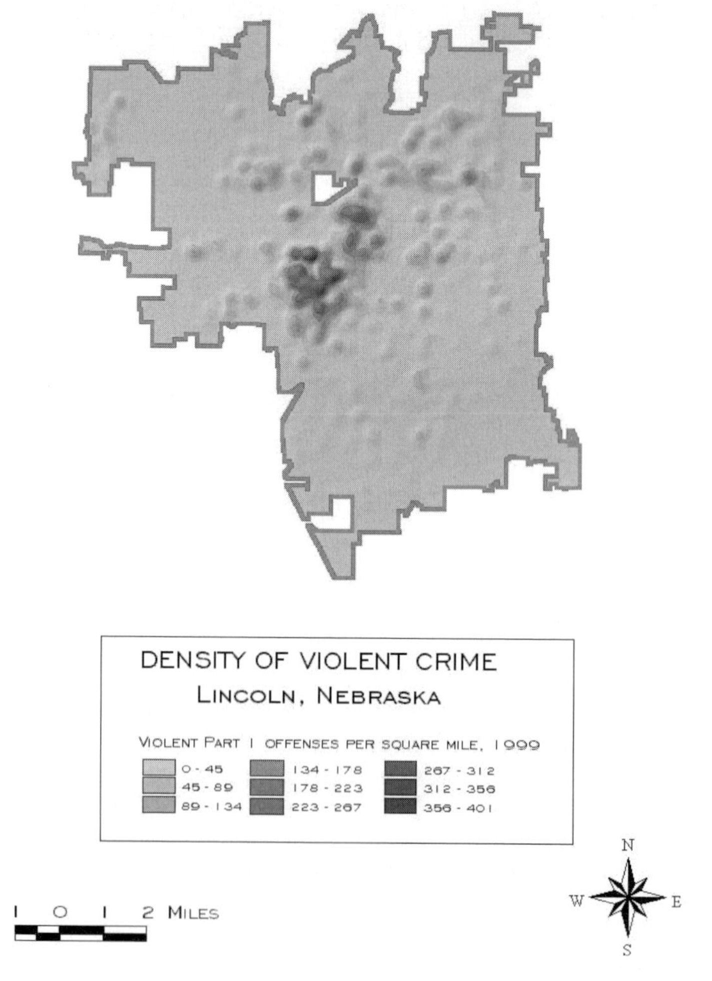

Figure 8.6 Supporting community development in fragile neighborhoods. CrimeView®

the F Street Recreation Center is well underway. Architects are preserving historic elements within the new campus design. The building will include a new substation for the LPD's Southwest Community Police Team.

Apprehending a flasher

During early 2000 GIS analysis revealed a series of indecent exposures occurring in south Lincoln (Figure 8.7). Detective Mark Domangue pinpointed the relationship between the cases based upon the suspect description, modus operandi, and geographic location. Despite a similar suspect description and MO, it is questionable that the cases would have been connected without GIS data, since they occurred over a several-months period and were up to two miles apart. As part of the department's regular analysis of crime patterns, however, the connecting thread became quite apparent: each of the cases had occurred near a recreational bicycle path. Although only a few actually occurred on the path, it was apparent that the flasher was using the trail as his travel route. After investigating one of the cases on 27 May, Officer Rob Brenner was conducting a neighborhood canvass. While doing so, he contacted an area resident who resembled the description of the suspect. Officer Brenner discovered that this resident had a previous record of similar offenses. Further follow-up by officers and detectives led to the arrest of Jason Callicoatt, who was charged with 13 counts of indecent exposure, and was later convicted on several of these offenses on 28 June 2000. Bicycle paths, rallies, greenbelts, flood control channels, and other non-road means of travel or ingress and egress are often worth mapping since in many cases criminals may use these corridors and not solely rely on streets and highways.

Keeping the public informed

LPD is one of a handful of police agencies that makes available interactive near real-time crime mapping on the Internet (Figure 8.8). The department deployed an interactive web mapping application in April 1999. To protect the confidentiality of victims, all data fields except case number and date are redacted. The application includes links to more detailed information about interpreting maps and understanding geo-coding. The department makes available more extensive tabular data about significant police incidents at the neighborhood level. This application lets citizens navigate to their own neighborhood and view a detailed table of incidents within the past sixty days. Together, these applications are generating 50,000–60,000 hits monthly on the police department's website. Both applications employ scripts and routines that automate their daily update, requiring only occasional and minimal attention from the staff. Lincoln's daily newspaper, the *Lincoln Journal Star*, began publishing simplified police crime maps in 1999. The Journal Star presently publishes maps three times each week.

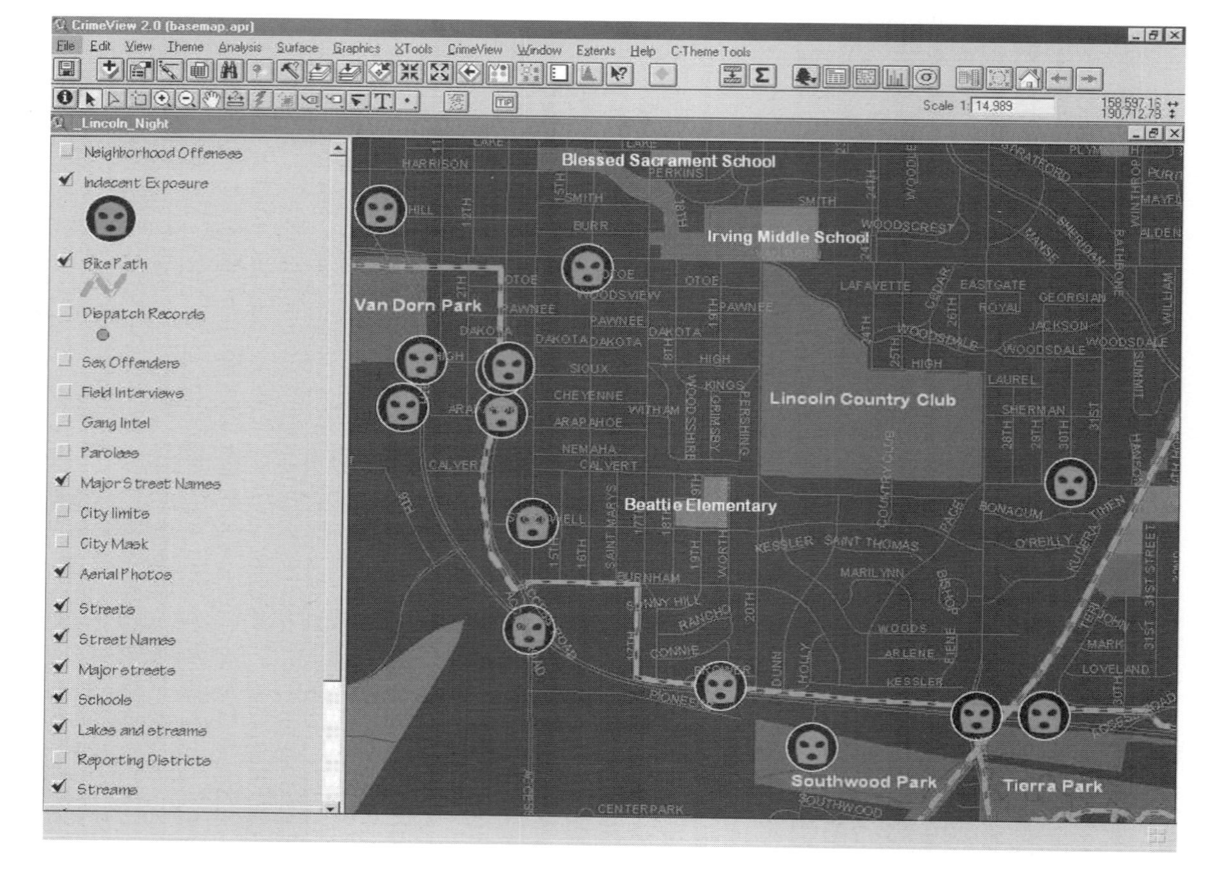

Figure 8.7 Apprehending a flasher. CrimeView®

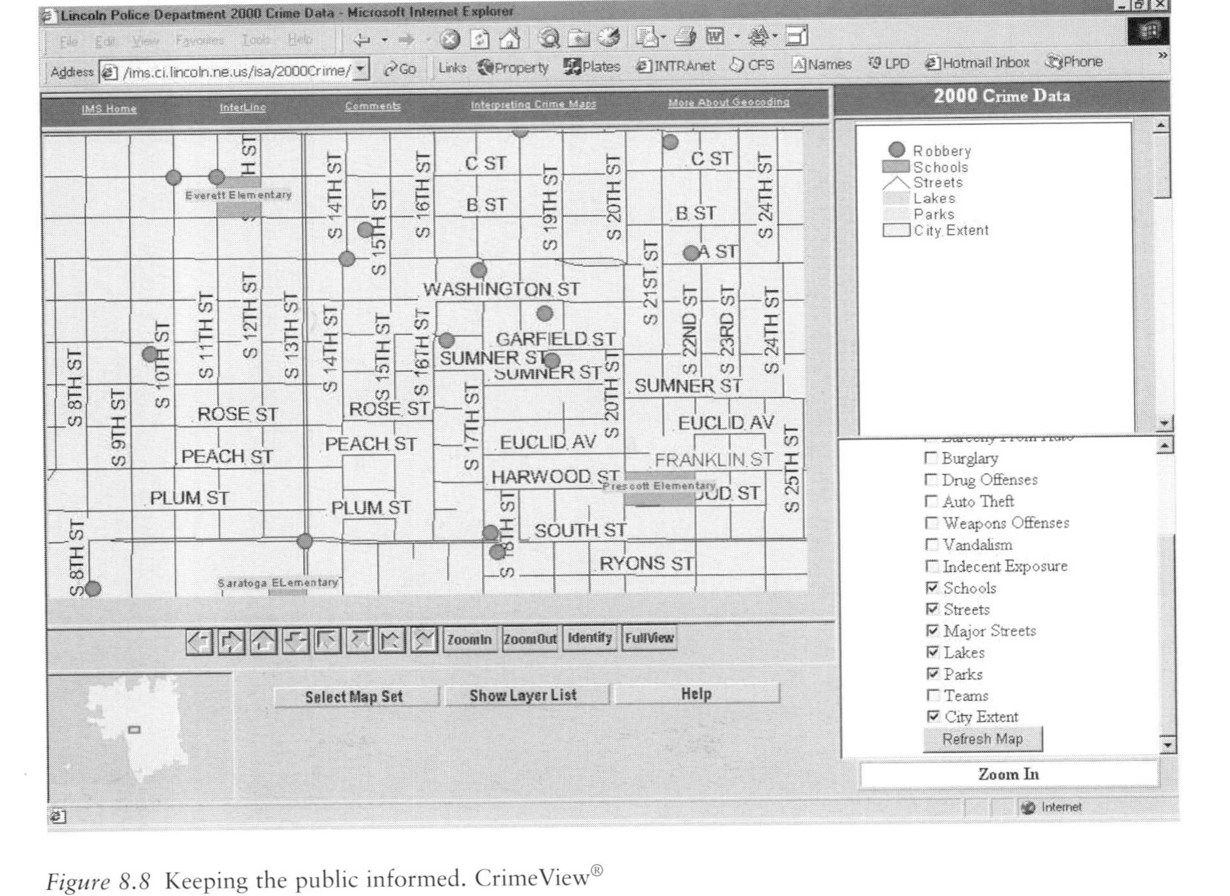

Figure 8.8 Keeping the public informed. CrimeView®

These maps generate considerable public interest and comment. Overall, Lincoln residents are considerably more informed about the nature and extent of crime in the community because of GIS. Keeping the public well informed energizes prevention strategies such as neighborhood watch, increases public support for law enforcement, and encourages tips and information from the general public.

9 Mapping crime and community problems in Knoxville, Tennessee

Robert Hubbs

In these cases from Knoxville, Tennessee the reader can see how GIS and crime analysis using GIS, has helped improve the efficiency of the community-oriented policing strategy adopted by the Knoxville Police. One of the cases presented also illustrates the use of GIS in apprehending a specific criminal, a serial rapist who preyed on young women in a spatially limited area. A common question about GIS in policing is if it can ever help apprehend a specific criminal: proof that it can is provided here.

(Editor)

If a child has been abducted, how would a police officer know where sex offenders live in the area of the abduction? What patterns of robbery are occurring on a patrol beat? A citizen is complaining about gunfire, noise, and speeding cars; what tools can be used to address these complaints? Traditionally, police officers have relied on reams of printouts and text lists of information. Now crime mapping is replacing lists, making it easier to solve crime and enhance community policing. GIS-based crime mapping is a great way to visualize overall community problems.

Regardless of the community size, crime mapping works well. For example, the Crime Analysis Unit, in Knoxville, Tennessee, uses mapping for crime analysis and offender tracking. It is a six-person unit serving approximately 450 police officers. It processes requests for crime maps, and tactical information for patrol, as well as citizen requests for neighborhood crime information. Knoxville is located in eastern Tennessee, with a population of about 170,000 and a daytime population of about 240,000 with 100 square miles within the city limits.

In the early 1990s Knoxville Police Chief Phil Keith made a commitment to use mapping for crime analysis and as a tool for police officers to visualize and solve problems. He saw how police departments in Los Angeles, San Diego, and Baltimore were using it. Realizing that mapping technology would facilitate community policing and problem solving, he upgraded the crime analysis capabilities of the police department.

It just makes sense to overlay all the police related data on a map to see how it fits. It is an intuitive tool, used by groups ranging from neighborhood watch organizers and community planners, to police officers. GIS is here to stay. Much like the police radio revolutionizing police communications in the 1920s, another information revolution is upon us in the form of information technologies such as GIS-based crime mapping and offender databases. Very rapidly, it is migrating onto laptop computers and hand-held devices, and soon will be widely used by officers in the field.

Mapping quality of life complaints – understanding crime and disorder

The problem solving examples in this section illustrate how crime mapping applies to the four steps of the SARA problem-solving model. These examples also illustrate how crime mapping is used to analyze data, track offenders, and to see the relationships between various pieces of information.

The SARA model

SARA is a method of policing that encompasses *s*canning, *a*nalysis, *r*esponse, and *a*ssessment. The main elements of SARA are described below.

Scanning is usually defined as observation of more than two incidents of a similar nature that are linked by time, location, type of crime or disorder.

Analysis consists of evaluating who, what, when, where, and why (suspects, victims, and locations). Patterns of incidents require analysis and problems rarely develop overnight. A crime triangle is a conceptual model in which victim, offender, and opportunity all play a role and must come together for a crime incident to occur.

Response is a tactical action plan involving directed patrols and perhaps community-oriented policing.

Assessment is the post-facto assessment of issues. It allows the officer to re-examine a problem and evaluate various responses and the effectiveness of analytical methods.

Typical crime analysis request

This case begins with the officer tasked to investigate a formal complaint. Over a period of a few weeks, several residents called the police department complaining of speeding cars, loud car stereos, and disturbances in their neighborhood (Figure 9.1). The complaints were passed down the chain of command to the beat officer to solve. Traditionally, the officer may try to use radar to interdict speeders and thus control the problem or simply try to do something like being present in the neighborhood in a patrol car when not answering other calls. Mainly, this type of complaint is viewed strictly as a "traffic problem."

However, police officers are trained in community-oriented policing to think about the root causes of crime and disorder. In this particular case,

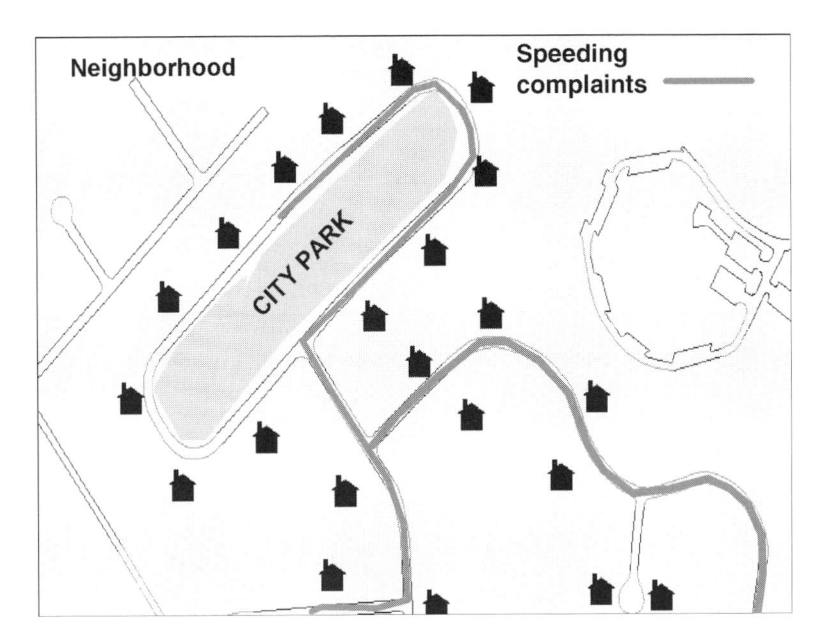

Figure 9.1 Speeding complaints.

the officer remembers answering several calls in the neighborhood, and the speeding complaints make him suspicious that other crimes besides speeding are involved. As part of a directed patrol process the officer is required to check with the Crime Analysis Unit for a neighborhood profile.[1]

In this case, layers of crime and other data such as 911 calls for service, reported crimes, parolee residential locations, city streets, and curbs are brought together within a GIS. Some of these databases belong to the Knoxville Police Department, while others are supplied by outside sources. For example, Emergency 911 supplies the calls for service data and the Tennessee Department of Probation and Parole supplies the parolee data.

What better way to understand the problem than visually? The crime analyst uses a drawing tool to define the areas of speeding and noise complaints. Mapping will show all the components of the problem. The sum total of information placed together on the map is greater than the sum of the individual parts.

Beginning the process

The crime analyst prepares a 911 report for the past 3 months in the grid that encompasses the neighborhood (Table 9.1). The analyst compares date ranges to see if there is an increase in calls for service. In our example, calls for service have increased concerning noise, fights, etc. This tabular report

Table 9.1 Example of a 911 report for a 3-month period

Grid 100	*Call type*	*1st month*	*2nd month*	*3rd month*
	Shooting complaints	0	2	3
	Disturbances	0	2	5
	Fights	0	3	4
	Traffic complaints	2	8	12
	Suspicious persons	2	6	14
	Vehicle burglary	1	4	5

Figure 9.2 Juxtaposition of speeding complaints, shots fired, drug parolees, fights, disturbances, and auto burglaries.

format shows the various categories of calls. However, our report still leaves us with more questions than answers. Tabular printouts are all that many police departments use for analysis, and prior to GIS all that was generally available.

Mapping the data will show that in addition to speeders and noise, there have been other calls to E-911 about fights, shots fired in the area, auto burglaries, and people under the influence of drugs and alcohol. Using coordinates supplied with the 911 data, a quick map can be made showing the locations (Figure 9.2). Interestingly, none of the callers mentioned these other problems in their speeding complaints. Our experience is that noise and speeding cars sometimes generate more complaints than incidents of a more criminal but perhaps less obvious character.

The crime analyst seeing the problems on the printout, creates a map of all the data. Fortunately, in Knoxville the coordinates (*X*, *Y*'s) are part of the 911 database record. This makes geo-coding simple. The analyst runs a query on the database and creates points on the map. Most 911 systems have mapping or address matching at the call taker level. Usually a call taker uses a computerized address lookup to pinpoint the address before it is sent to the dispatcher. This is something worth checking on in your community since it will be a major windfall for your mapping efforts if you are getting started.

Our mapping process in ArcView GIS from ESRI

1 Using a polyline tool for drawing the speeding complaints.
2 Creating points from our *911 calls* for service database using supplied *X*, *Y* coordinates.
3 Mapping reported *crimes* from the crime data table.
4 Layering *parolees'* and probationers' last known home addresses on top of all this.
5 Adding curb lines (rather than street centerlines) and building footprints.

A flow chart of spatial events is developing

Notice that most shooting calls originate from a two-block area near the back of the neighborhood's cul-de-sac. In addition, disturbances, and fights symbolized by the hammer and stick figures are all in close proximity to one another. Other surprising developments begin to emerge as the crime and parolee data layers are added. The burglaries, mostly from cars parked in driveways occur overnight, especially on the weekends.

The problem is more than speeding cars!

The maps (Figures 9.1–9.2) show that the pathways that lead into the neighborhood are the same areas prone for the increase in auto burglaries as well as the locations of the speeding complaints. As the map layers are further placed over the neighborhood a symbol indicating a drug parolee's address appears near the end of the cul-de-sac. What a coincidence! This is the same area that gunshots had anonymously been called in to 911 several weeks prior to the analysis. In this case, the parolee's criminal history included drug addiction, drug trafficking, and property crimes.

Armed with this information an officer went to the Narcotics Unit to ask about any complaints that had been logged on the drug hotline. As expected, the narcotics investigators were aware that the parolee probably once again was engaged in drug dealing. However, when they saw the map they knew that an immediate investigation was warranted. The investigation cumulated in a search of the home, and arrest of the parolee. It was later learned that many of the parolee's customers had been involved in behavior that prompted the disturbance calls and had engaged in crimes both inside and outside of the neighborhood.

Mapping brings it together

Traditional policing must be contrasted with a problem-solving orientation. Often, officers believe that more patrol will solve a problem; that through sheer presence of officers the criminals will leave an area. While this may provide a short-term solution it only displaces the problems and the police can't be everywhere at once. This case, in the past, would have been viewed strictly as a "*traffic problem*" with no analysis and no coordination with other units, *especially narcotics*. Essentially, one map brought together various units in the department, and paved the way for the solution of the problem. Many officers had been aware of the offender, but the map took institutional knowledge, and added it to the street-level experience that already existed to help officers better interpret what they were sensing. Of course this crime mapping problem-solving model depends on officer training, experience, and data. Last but not least, seize the opportunity to follow-up with the community after solving the problem. Provide feedback to neighbors that reported speeding cars and they might be willing to report other incidents. This is a great opportunity to start or enhance a block or neighborhood watch program.

Tracking parolees and serious habitual offenders

Research conducted by criminologists (Brantingham and Brantingham 1984) suggests that most property crimes are committed within two miles of the offender's residence. Mapping allows us to visualize this and develop response strategies from this fact. There is a good likelihood that someone fresh out of prison on parole will repeat some of the crimes that put him or her in prison in the first place. Maps showing the offender type, previous charges, race, and sex provide an avenue to track parolees (Figure 9.3). These can all be symbolized on the map as long as the data are available in the appropriate format (i.e. geo-coded) to do so. Also, queries to find specific tattoos or the vehicle type owned by a parolee can be run to make a custom map. This works even if the offender's face was covered in an armed robbery and all the witness saw was a tattoo, mark, or car make. Many times investigators want maps showing the parolees convicted for armed robbery or a specific physical characteristic. This process offers investigators a starting point.

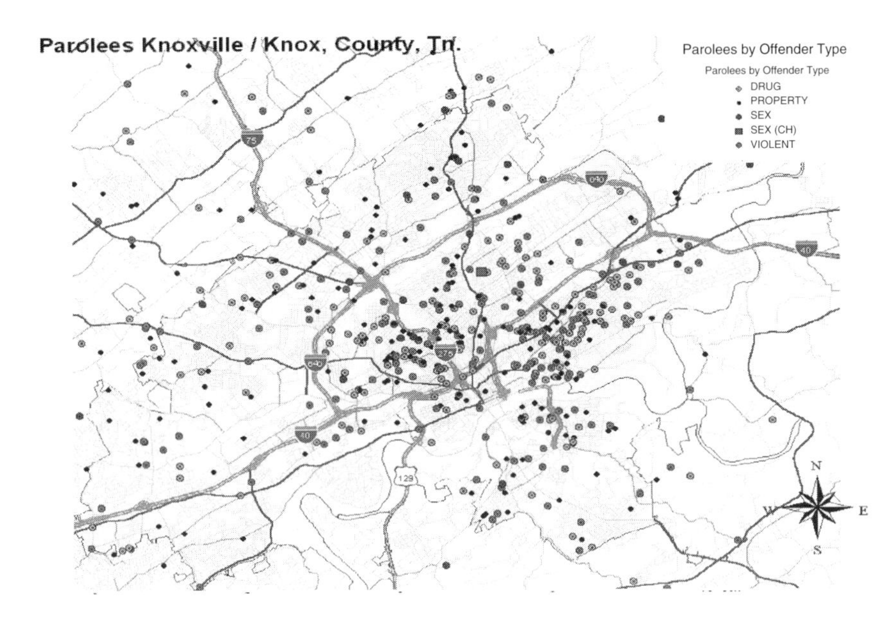

Figure 9.3 Parolees by offender type, Knox, County, Tennessee.

Anyone visiting a police roll-call room or detective bureau has probably seen large clipboards containing lists of names and individual release sheets on parolees. Computerization of this data including GIS-based mapping of parolees over a city map is a much better way to begin tracking these individuals as they are released from prison.

Parole I.D. card/psychological deterrent/crime mapping tool

A simple and effective way to track parolees was developed in 1995 in Knoxville, Tennessee. Police Chief Keith and Jim Cosby, Tennessee Director of the Department of Paroles and Probation, decided they would develop a simple identification card color-coded by the type of conviction. This process begins with the newly released parolee reporting to the police department for fingerprinting. The parole board issues them a photo ID (Figure 9.4). Maps are made from a database sent by the parole board each month containing current address, description, tattoos, vehicle description, and electronic photo attached. Should parolees encounter a police officer for any reason they are required as a condition of their parole to show the identification card. Failure to do so will result in punitive administrative action that could include parole revocation.

A query of parole names is run that compares them to all police field interviews, victims, complainants, suspects, and pawn information. This is a valuable tool because it is sent to the parole officer. When the parolee

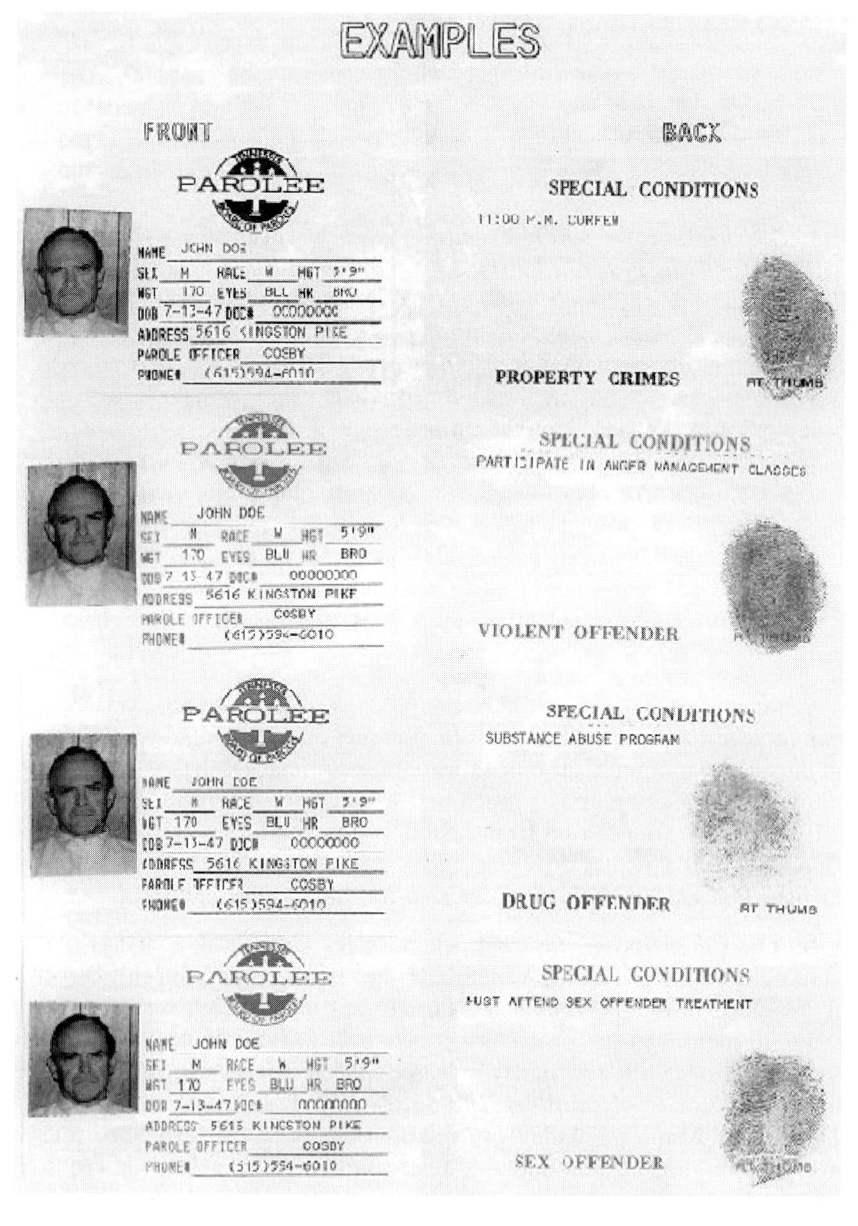

Figure 9.4 Examples of parolee photo ID cards.

reports each month the parole officer can ask about an auto accident or field interview or anything that the parolee may have been involved in. This is a way to monitor parolee activities and contacts with the police department and to verify compliance.

This has moved us from a single sheet being mailed to us to a comprehensive data table and associated geographic features stored in GIS layers almost overnight. The parolee card is a great psychological tool that lets the parolees know that they need to behave because they are being monitored.

Why map sex offender addresses?

The headline "Ex-con held in rape of 8-year-old" *exemplifies* the need for tracking sex offenders and parolees in your community. Sadly, headlines like these can be found almost anywhere in the United States today, and Knoxville, Tennessee, is no exception. The incident occurred in 1993 at a time when no mapping and tracking ability for known offenders existed in the city. Desktop mapping was still new to law enforcement, and while there were plans to use it in the future, no system was available that geographically tracked known offenders. Much like the Polly Klaus case in California, our victim was abducted from her bedroom overnight. Unlike the Polly Klaus case, our victim was found alive. This case underscored the need to be able to locate and determine exactly who is out on parole and where they were last living.

Dr Janet Warren, a researcher with the Institute of Law Psychiatry, suggests that some serial rapists might live as close as 3 miles of their attack scenes (San Francisco Chronicle 1997, A1). And research from the National Center of the Analysis of Violent Crime suggests that rapists attack outward in a "V" shape spreading from their home neighborhood. How will you see this pattern if you are not mapping?

Maps make lists obsolete

In policing we have all types of lists for known offenders, parolees, sex offenders, warrants, gang members, etc. But how is one supposed to take a list, especially a large one, and figure out who lives on your beat, or which offender lives near a crime scene. Most large lists are sorted *alphabetically* by the last name. An alphabetical list is great, if one knows whom to look for. Most of the time we don't. What if one needs to see all the people on the list in proximity to a crime scene? Long lists of names sorted alphabetically do little to aid in the geographical analysis of offenders and of their relationship to crime (Figure 9.5).

In Tennessee, sex offenders must register with the Tennessee Bureau of Investigation. An alphabetical list by offender last name is sent quarterly to area police departments in the form of a greenbar printout.

However, without spatial or geographical attachment of this data to a map, the investigative potential of the list is not fully realized. The list is useful in its tabular format but it is difficult for an officer to "see" if any of the over 200 offenders on the list present in Knoxville were living near a particular crime scene or near a school or other target for sex offenders.

Figure 9.5 Long lists are cumbersome.

For that reason, we have created a database and keypunched the data on the list into the database and mapped it (Figure 9.6).

Serial rapist caught with mapping

In 1997, a serial rapist was terrorizing women using the Third Creek Greenway in Knoxville, Tennessee (Figure 9.7). This case was solved by mapping known sex offenders and comparing them to crime locations.

On 27 August 1997, a victim was raped who had been walking along the Third Creek Greenway commonly used as a running and biking recreational trail. The victim was very traumatized, waiting about eight days to report the incident. This delay somewhat hindered the investigation. Investigators used a female decoy officer on the trail and patrols were increased but nothing further materialized. About ten days later on 16th

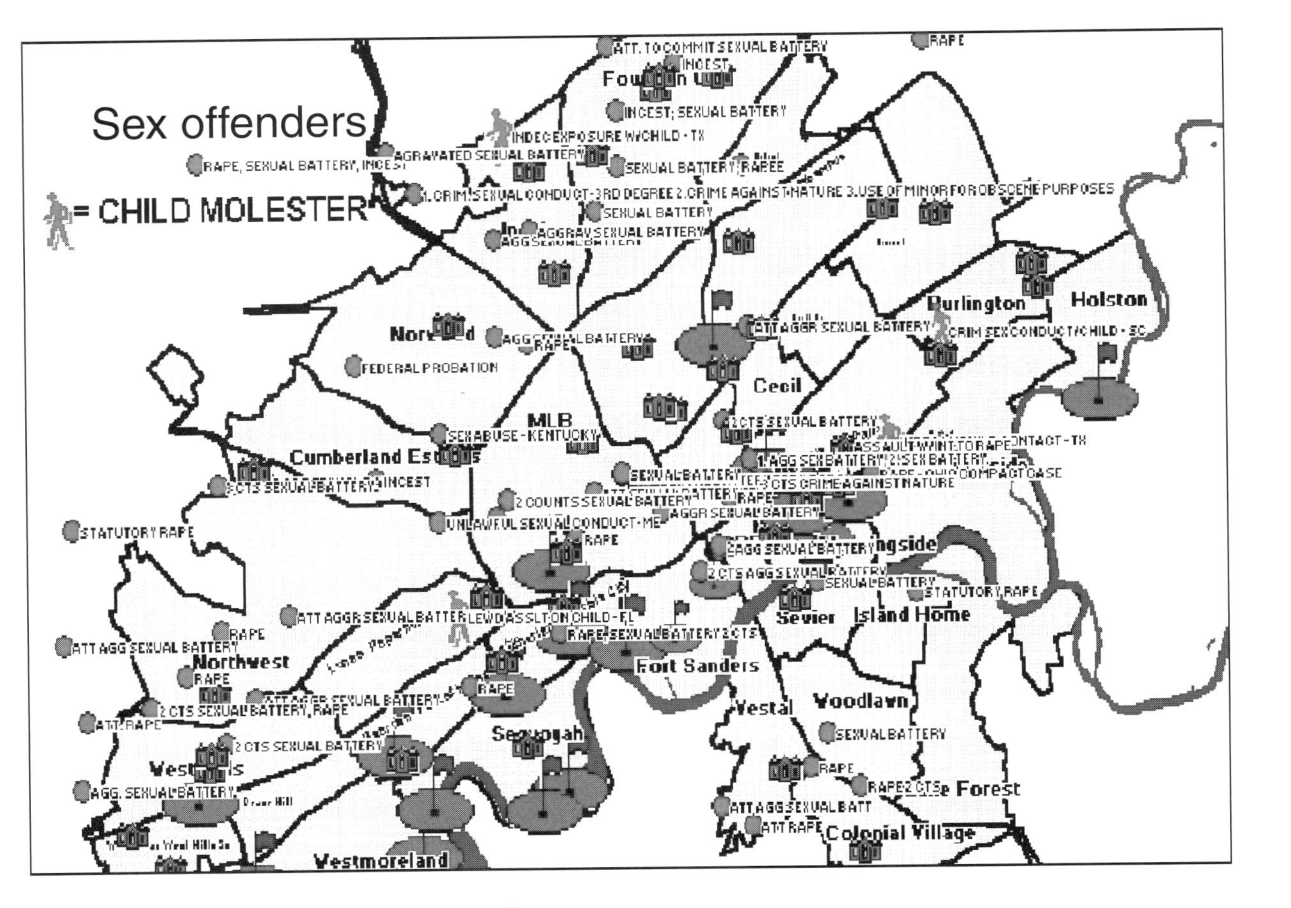

Figure 9.6 Location of sex offenders, Knoxville, Tennessee.

Figure 9.7 Third Creek Greenway, Knoxville, Tennessee.

and 17th September two rape attempts were committed by, most likely, by the same perpetrator.

Investigator Tom Pressley was assigned to help with the next two attacks. His first call was to the Crime Analysis Unit, and he requested that a map be made of sex offenders, parolees, and serious juvenile offenders who lived near the crime scene. He also requested a list of names and descriptions. A map of those living within a two-mile radius was prepared of offenders using the databases supplied by the Tennessee Board of Paroles, Tennessee Bureau of Investigation (sex offender data), and the Knoxville Police Department's Serious Habitual Juvenile Offender Program (SHOCAP) database. Also a list showing their description and their offender profiles was produced.

The solution found on a map

The victims of the last attacks provided better descriptions. A map and the accompanying data helped narrow the focus of the investigation. Some of the offenders were closer matches because of their physical description while others were ruled out completely.

It was easy to see that there were several sex offenders living within half a mile of the crime scene. But without the spatial analysis of the three databases layered on top of the crime scene this information would not be readily known.

Investigators took information from the map and made photo lineups of suspects fitting the rapist's description. On the first photo lineup both victims picked a sex offender living nearest the trail as their attacker. He was confronted by investigators and confessed to all three attacks and one unreported attack. He, eventually, was convicted of all these crimes. GIS has also been used to nab serial rapists in Las Vegas, Nevada, Detroit, Michigan and Austin, Texas.

Other community policing applications

Figure 9.8 shows the spatial distribution of neighborhood watch zones and the location of watch coordinators within Knoxville. Knowing the "holes" in coverage can be a useful instrument to solicit participation for a neighborhood watch program.

Gun crime study

About 4 percent of the City of Knoxville, Tennessee, 100 square miles area, accounts for over 50 percent of shooting calls, murders, and other violent crimes. These findings would also be similar if this study were conducted in many other mid-sized and large US cities. Before one can work on the problems one needs to put them into perspective.

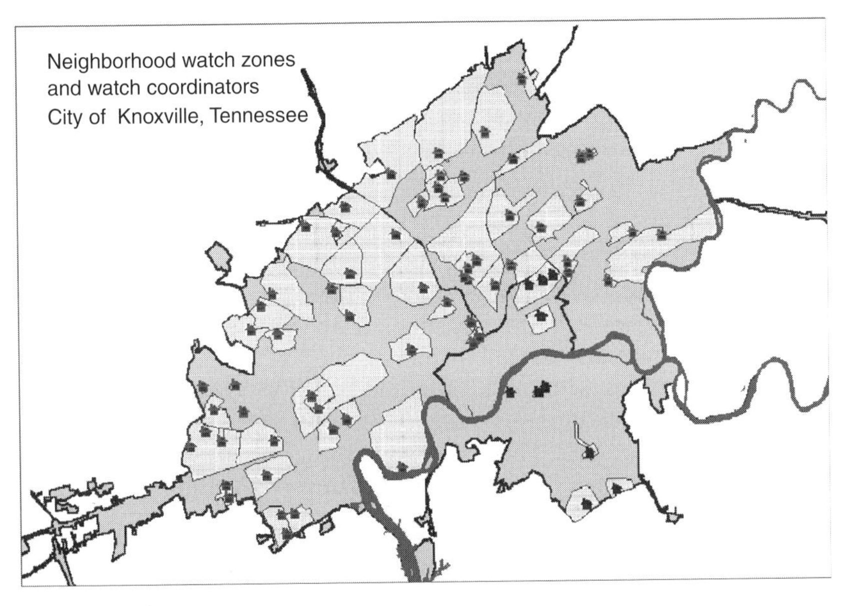

Figure 9.8 Neighborhood watch zones and watch coordinators, Knoxville, Tennessee.

The objective of the gun crime study was to analyze locations of firearm violence occurring in Knoxville, Tennessee, and to determine the areas having the highest concentration of crimes committed by use of a firearm.

Empirical data suggests that certain areas of the city are hot spots of gun violence. However, the actual boundaries or "gun zones" were never defined, and have been subjectively turned into "mental maps" based on the experience of the individual beat officers or investigators.

Officers, when interviewed, maintain that for many years small areas of the city they call "gun zones" are subjected to nightly sounds of gunfire and gun violence. They liken this to a form of domestic terrorism. Many residents in the "gun zones" hear nightly gunfire and see the evidence of its aftermath the following morning with the addition of new bullet holes in parked vehicles, street signs, and buildings. According to officers the perpetrators may not even live in the affected neighborhood but come to sell or purchase illegal drugs or may be involved in other illegal activities.

This study sought to validate with data and precisely define what beat officers already know. Officers know that certain parts of the City are more violent than others and that gun use and violence has increased as a result of the presence of drug dealers and gangs, and that illegal gun use and drug markets overlap.

Beyond validating known "gun zones," there is a need to reduce homicide and gun-related violent crime in these zones. Another rationale for conducting this study is to differentiate between emerging "gun zones" and established "gun zones" because the tactics likely to be employed and impact of the tactics will be different in each area. Emerging "gun zones" are in a state of transition and will be more responsive to applied countermeasures and maintenance. Established "gun zones" will possibly be less responsive to applied countermeasures and will require more maintenance. Lastly, it is hoped that this study can be used as a model for other cities with similar problems. Arming police officers with tactical data is as important as any other tactical tool used for suppression of crime.

Where does most gun violence occur in the city?

Data was collected on shootings that occurred in Knoxville from 1995 through 1998 (Figure 9.9). The locations of the shootings were geo-coded, mapped, and thematically shaded by traffic zone.[2] In the maps below, darkest colors indicates zones with the highest number of shootings, while progressively lighter fills indicate lesser number of shootings with the white areas having the fewest incidents. The maps clearly show the City's traffic zones that have historically reported the highest number of shooting calls for service.[3]

Cluster analysis

Using Spatial Analyst, an ArcView mapping extension, point cluster analysis on shooting calls helped to determine "hot spots."[4] "Hot spots" are

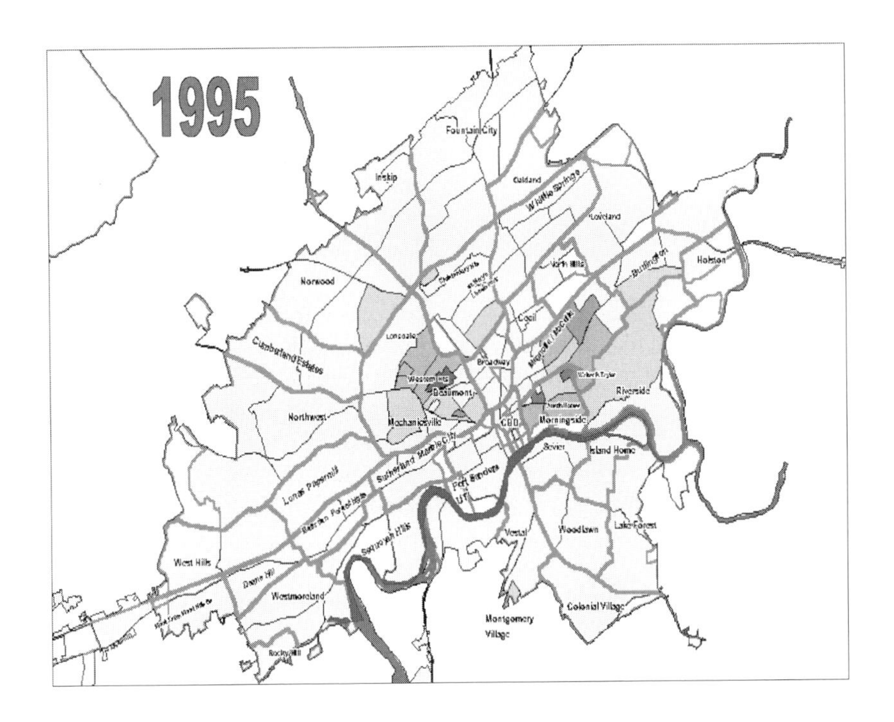

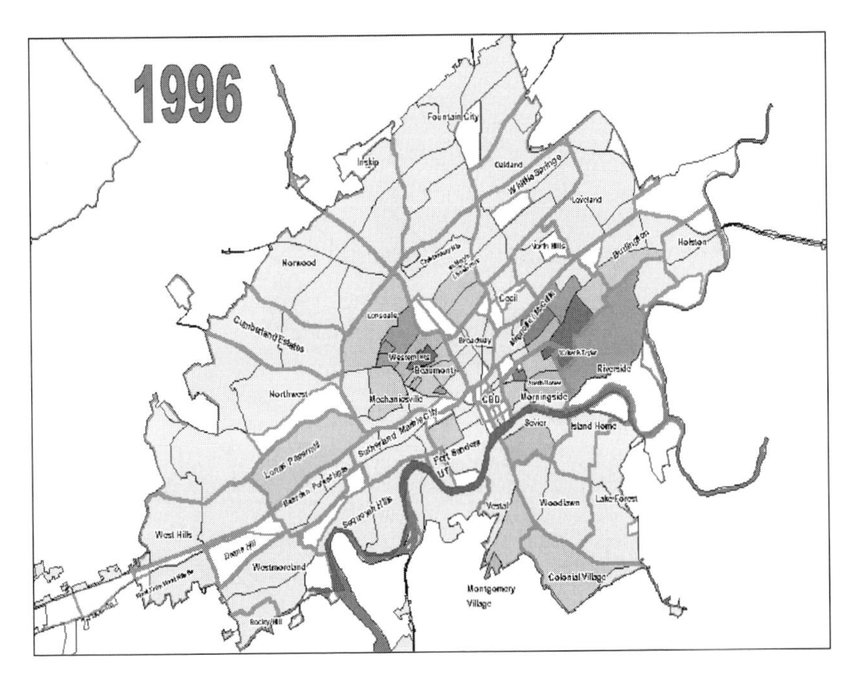

Figure 9.9 Location of shooting by traffic zones, Knoxville, Tennessee, 1995–98.

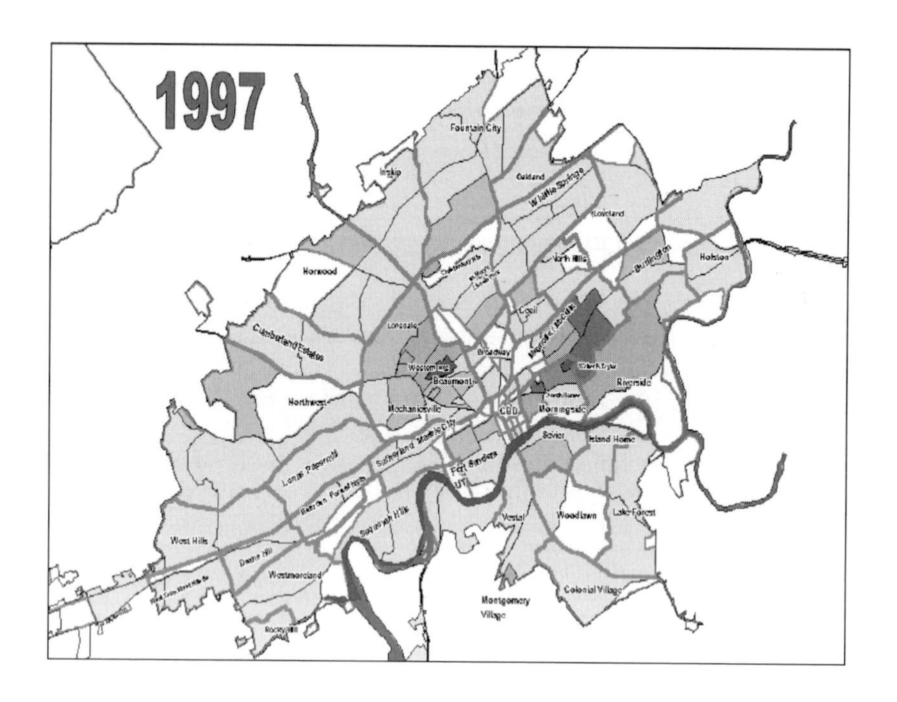

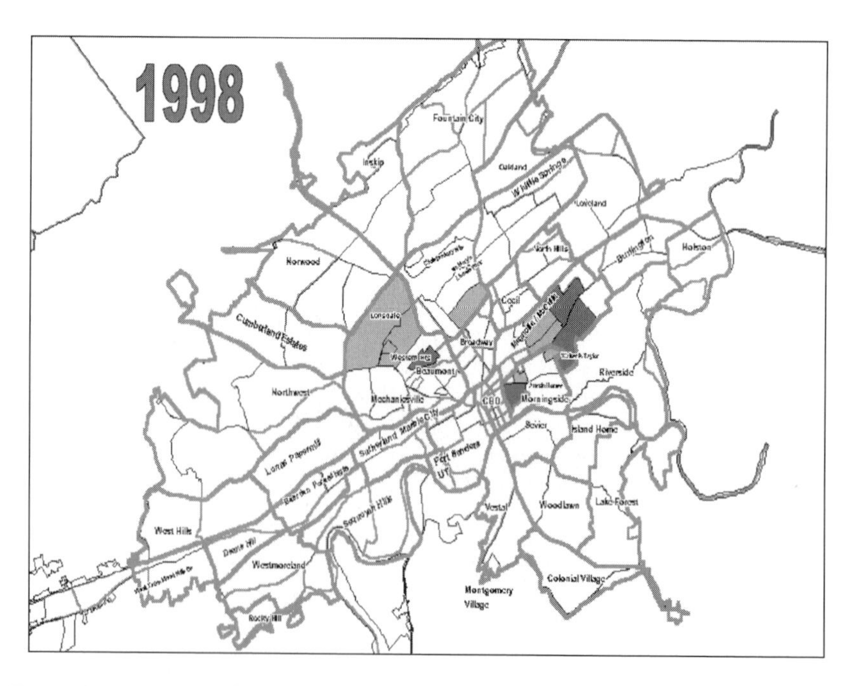

Figure 9.9 (Continued).

defined by electronically girding the city into 250 square grids and using a 1500-foot radius to calculate the next nearest neighbor.

Shooting call hot spots from January through August 1999 were analyzed (Figure 9.10). The results were consistent with that of the thematically shaded traffic zone maps from 1991 through 1998. However, using Spatial Analyst, the areas plotted in darkest fills are more location specific, showing a clustering of points where shooting calls have originated. The "hot spot" intensity with the darkest color indicates highest density. Not only were hot spots discovered from a multitude of point locations, but emerging hot spots were discovered that were consistent with community complaints about two crack dealers who had set up operation in the neighborhood. One of these offenders is currently in federal detention pending charges of gun possession while dealing crack cocaine. This map was used in the bond hearing. This is a good example of GIS-based crime maps not only helping to arrest offenders but being used in Court to make sure they stay behind bars.

How much violent crime occurs in the identified gun zones?

The identified gun zones are approximately 4 percent of the City's area and much of the City's violent and gun-related crime occurs in the two identified gun zones. Also "masked" gunfire incidents were found and added. This is the case where the report is written up as vandalism but in reality bullets from a gunfight have entered an occupied dwelling damaging it or

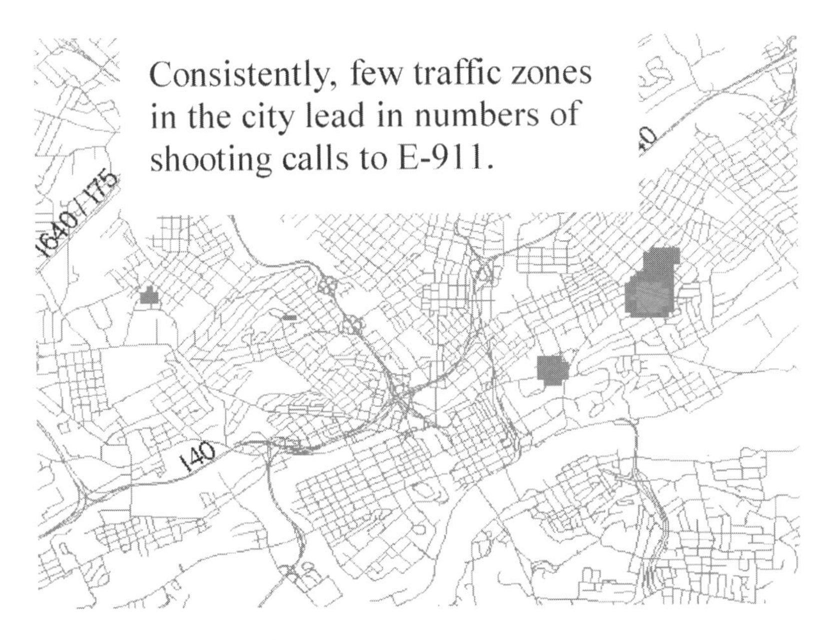

Figure 9.10 Hot spots of shooting calls to E-911.

bullets have struck parked cars. During 1998, the following violent and gun-related crimes occurred in the identified gun zones.

- 54 percent of the City's murders
- 47 percent of street robberies with a firearm
- 51 percent of "shots fired" calls
- 73 percent of aggravated assaults with firearms on the street
- 47 percent of total guns confiscated.

What countermeasures might prove effective?

- Plan and conduct saturation patrols of open-air drug markets.
- Educate police officers on the elements of federal firearms violations (922 G and 924 C).
- Increase training protocols for handgun seizure evidence.
- Work with the Bureau of Alcohol, Tobacco, and Firearms Taskforce to ensure criminals arrested in the gun zones, qualifying for federal sentencing for carrying/using a firearm in the commission of a crime are prosecuted.
- Configure the county warrant computer to cross-check parolees, career criminals, and sex offenders.
- Advise the Attorney General and Federal Prosecutors of the study.
- Identify, and arrest career criminals known to go armed in the identified gun zones.
- Ensure that warrants for violent/gun offenses in the gun zones are served promptly.
- Present this study to the Board of Probation and Paroles.
- Track activity of parolees and probationers who live in or frequent the gun zones.
- Plan and conduct proactive measures to curb illegal gun sales.
- Survey the community to collect data regarding community perception of the problem.
- Work with community groups to develop support and solicit ideas for shutting down the gun zones.
- Work with community groups to monitor sentencing for offenses committed in the gun zones.
- Communicate gun zone study findings to the City Council and solicit ideas to solve the problem.
- Communicate gun zone study findings to the media and request community support.
- Run cross-check queries on known offenders who have been documented frequenting these areas.

Conclusion

It just makes sense to selectively overlay crime data on a map to see how it fits. It is an intuitive tool for analysis. Almost 500 years ago Reneé Descartes, a fifteenth-century philosopher and mathematician offered this advice:

1 Accept as true only what is apprehended so clearly and distinctly that you cannot doubt it.
2 Break up each problem into as many parts as it will yield, tackle these in turn.
3 Observe and order your inquiry, passing from the simple to complex, from what is easy to understand to what is more difficult.
4 Make sure of covering the whole ground (Allen 1992).

Does this sound familiar? Of course it does, but what better way of doing all of this but with a map!

Notes

1 Directed Patrols or DPs require officers to write the problem down and do an assessment using data.
2 The City is divided into geographic areas called "*traffic zones*." Traffic zones generally fit neighborhood boundaries and fit into census tracts. Shooting calls were totaled by traffic zone and shaded (colored) according to the number of E-911 calls in each zone. This process is called "thematic shading."
3 Addresses for shootings reported at hospitals were removed from the data as they generated numerous calls for officer investigation of "walk-in shootings" but did not reflect an address where the shooting actually occurred.
4 ESRI's ARCView extension *Spatial Analyst uses* mathematical algorithms to perform point-cluster and density analysis and determine existing and emerging "hot spots."

References

Allen, E. L. (1992) *From Plato to Nietzsche*. New York: Fawcett Premier.
Brantingham, P. L. and P. J. Brantingham (1984) *Patterns in crime*. New York: Macmillan.
Lee, H. K. (1997) Investigators scramble to profile rapist: Cops using psychology, history in hunt for East Bay attacker. *San Francisco Chronicle*, 25 Monday, A1.

10 Operationalizing GIS to investigate serial robberies in Phoenix, Arizona

Bryan Hill

GIS can be applied by a crime analyst to general issues like of the distribution of crime in a community or to a specific series of interrelated crimes. Given limited resources, these serial crimes under analysis using GIS are most commonly serial murders or rapes. However, in this case from the largest city in the fast-growing and diverse state of Arizona comes an application to serial robbery.

(Editor)

There are 2,600 sworn officers serving the 1.2 million people of Phoenix, Arizona. Personnel assigned to the Phoenix Police Department have developed several strategies to do their jobs better over the past twenty years. Many of these strategies are currently being used throughout the United States and are modeled after programs that began in Phoenix, Arizona. The PPD has also made use of programs created by other dedicated officers from other jurisdictions and continues to find and promote new ideas to combat old problems. Technological advances in the past twenty years have had a great impact on the way police departments do business. The Phoenix Police Department has a long history of being on the front line of technological advances in policing. PPD makes daily use of technologies such as the automated fingerprint identification system (AFIS), police automated computer entry system (PACE), computer-aided dispatch (CAD), and other products designed to assist the police in capturing and identifying suspects involved in crime. One of the "new" technologies being widely used throughout the department is geographic crime fighting, or geographic information systems (GIS) technology.

When a police officer first hears about the GIS technology he/she usually thinks of the paper pin map created for a community-based policing project they were involved in. The paper map on the wall was a way to track crimes as they occurred, but the pins weren't very smart and consistently fell out of their assigned places over time. GIS allows police officers and analysts to create "smart" pin maps that are electronic versions of those age-old wall maps. One of the leading reasons why GIS technology has started to grow in popularity in Phoenix is community-based policing itself. The community-based policing philosophy creates a dilemma for police agencies in that every problem identified is important; however there are only a limited amount of resources a police supervisor can dedicate to one problem before he/she moves on to the next problem. Traditional community-based policing strategies involve a technique dubbed the "SARA" model. This acronym stands for SCAN to identify problems, ANALYZE the problem and what would be needed to fix it, RESPOND to the issue identified and put your program into action, and ASSESS the success of your program. It is my belief that police departments in general do the S-A-R part of the SARA model very well. We can find and identify problems of an enforcement nature either through citizen requests or active participation and partnership with the police officers and the community. We can analyze what the problem is and send officers into an area of the city to arrest criminals engaged in problem causing activities. In the process we also partner with other city departments and private organizations in order to improve the quality of life in the city. Police departments evaluated their success by a report listing the number of arrests made, warrants served, calls for service reduced, or perhaps crack houses demolished with the owner's permission. However, there was still something lacking in our confrontations with crime and we found that there were different goals and objectives for various stakeholders in different parts of the City of Phoenix.

Then a new concept was presented called the "Policing Plan." The Policing Plan involved surveying citizens and Police Department Staff to find out what their perceptions of crime and quality of life were. This information was then compared and evaluated with statistical and other data, and a plan was developed that allowed everyone in the department to share several common goals. Now, the Chief, the Assistant Chiefs, the Commanders, Lieutenants, Sergeants, and individual officers had a common goal to work toward and a more focused effort could be developed. In any organization the size of the Phoenix Police Department, there will be some hurdles and challenges. However, the adoption of the Policing Plan made some of those hurdles and challenges a source of common effort. One of the aspects of the Policing Plan was to find a way to measure the success of employees at meeting the goals established. A quick glance at the rest of the United States identified an emerging trend in the form of the "COMSTAT" processes. Under a ComStat system, police commanders have to show crime statistics for their areas at a meeting with the Chief and be responsible, if there were increases or decreases in crime in their districts. The ComStat name comes from the program first developed by the New York City Police Department. GIS and crime mapping was viewed as a great tool for graphically displaying this information and statistical data. Administrators in the Phoenix Police Department wanted to use this new technology as well. Individual employees who now had personal goals and tasks and who were using the SARA model to identify a problem in their own areas caused a secondary impetus toward the adoption of GIS. They would also have to analyze the data, and create a program to meet their own goals. The demand for more accurate, up-to-date, and easily understood data and access to tools became a predominant thought for members of the force.

Employees from the PPD were able to attend several conferences, during this time period, that dealt with crime mapping and crime analysis efforts across the United States. The most valuable conference was in Denver, Colorado in December of 1997. The National Institute of Justice (NIJ) Crime Mapping Research Center (CMRC) brought presenters from all around the United States to discuss spatial analysis of crimes and criminals. For the first time, PPD staff were able to see the impact and use of GIS technology on crime-fighting efforts. They found out that there were a vast number of possibilities for a police department to not only identify problems, but also analyze data, make informed decisions on deployment issues, and evaluate the success of programs through a single tool. GIS also allowed for a method to bring information and data to the user and has proven to be a powerful tool for crime analysis.

The next step was to determine what *crime analysis* was. Over time, the Phoenix Police Department was able to purchase copies of ArcView and CrimeView and start developing procedures for the analysis of crime. Although this process is an ongoing and constantly evolving effort for the department, the PPD has been able to make great strides toward the goal of having a pro-active crime analysis unit in Phoenix with the assistance of the "common language" of GIS. Crime analysis definitions and experience were gained through training and trial and error. The capriciousness of human nature is a big factor in crime patterns and this impacts analysis efforts. Therefore, finding and applying theory to a set of crimes is not always operationally effective. Analysis of a series of robberies presented itself as the first problem we tackled effectively using GIS-based crime analysis techniques.

The Blue Bandit, Super Sonic Bandit, and Panty Hose Bandit robbery series

The "Blue Bandits" robbery series is the example of an issue to which the PPD crime analysis staff sought to apply GIS-based crime analysis (Figure 10.1). This involved a series of armed robberies of supermarkets primarily in Phoenix, Tempe, and Tucson, Arizona. We found that this series crossed several jurisdictional boundaries and may have even involved crimes in California. We actively geo-coded the crimes in Phoenix and tried to see if there were any obvious patterns to the robbers' attacks. We worked closely with the Robbery Detail and it should be noted that it was their idea to come to us for this GIS-based analysis effort when they noticed that a series of related crimes existed. We were able to make a prediction about *when* the next robbery might occur using basic crime analysis date and time statistics.

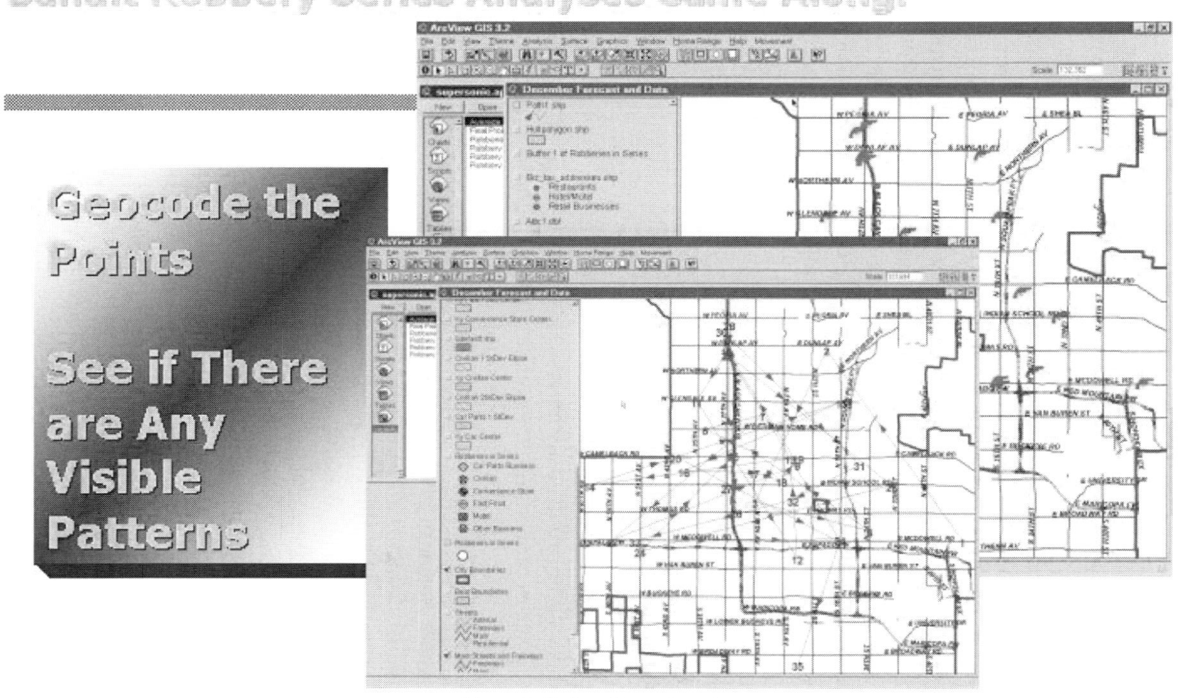
The Blue Bandit, Supersonic Bandit and Panty Hose Bandit Robbery Series Analyses Came Along!
Geocode the Points
See if There are Any Visible Patterns
ArcView GIS 3.2
December Forecast and Data
Phoenix Police Department –
Research & Analysis Unit

Figure 10.1

Our first goal was to see if we could apply other database information to the problem to help the Robbery Detail identify *who* a suspect might be. We used the traffic citation data, traffic crash data, field interrogations, and known offender data to try and come up with a name for the Robbery Detail. We used several methods to try and isolate the possible candidates down to a manageable size list. The outcome was still a list with hundreds of potential suspects and the operational use of the information was questionable. Another operational problem was being able to actually tell the Robbery Detail personnel *where* the next robbery might occur. Because of the wide-ranging activity area for these suspects, the potential area of the next robbery covered a region of twenty-five or more square miles. This encompassed most of the City of Phoenix and wasn't very useful preemptive information to the detectives. We went in search of a database of businesses for the City of Phoenix and eventually found an electronic version of a business listing and geo-coded those addresses. We significantly reduced the number of potential targets by creating a map of all of the grocery stores of the type these suspects seemed to exclusively hit. The contrariness of human nature then took over and the suspects never hit again in Phoenix. We could not improve our analysis process until months later when the "Super Sonic" bandit series was identified by the Robbery Detail and was brought to the attention of the crime analysis unit.

The Super Sonic Bandits committed thirty-five robberies until a suspect was arrested after a near-fatal shooting of an Auto Zone manager/victim. This robbery series allowed us to add to the techniques we learned in the Blue Bandit series and continue to make a more operationally reliable product. The process of identifying potential suspects through other databases had about the same usefulness as in the Blue Bandits series. We had too many possible suspects and not enough information about the suspect to whittle the list down to a more manageable list of potential suspects. In the geographic analysis of this series, we did see a pattern in the way the suspect was hitting targets. This was only evident once we had mapped the robberies and created a path from robbery to robbery. The pattern was that the suspect would rob a store in the Northeast corner of the City, and then the next hit would generally be in the Southwest corner. This went back and forth as he alternated back and forth across the city with the area of Bethany Home Road and the Freeway being the center point of the robberies. In addition, one of our analysts developed a method of using our quarter square mile grid boundary theme as a probability surface for the next robbery. His method, coupled with store locations allowed us to specify certain, more likely, businesses the suspect might strike next. During the later part of the robbery series, the suspect was seen by undercover units at one of the targets we had identified. However, he eluded the officer in traffic. We now had a more operationally useful map that we could provide to officers that showed potential targets, but we had no way to determine where the suspect might live to help focus investigative efforts in that sector. Geographic profiling is a relatively new process of determining where a suspect may reside or offend based on spatial factors and the locations of prior crimes. Using this method, it was predicted that the suspect would most likely live in the area of 19th Avenue and Bethany Home Road. The suspect was eventually caught through investigative efforts of the Robbery Detail, which had little to do with our crime analysis product. It was interesting to note however, that the suspect gave his address as a home near 12th Street and Southern, but he had been living out of motels near the intersection of 19th Avenue and the freeway near Bethany Home Road. This information only fueled our belief that we were then and are now on the right track with the use of GIS based geographic profiling techniques and that if we had a bit more information, we might be more effective in assisting detectives and officers "catch the bad guys!"

The Panty Hose Bandit series involved only four crimes and then abruptly ended. We applied all that we had learned in previous series analysis and believe our product to be very useful. However we still don't know what else we could have done, had we only known more!

The actual process

The steps involved in doing robbery series analysis are generally described as follows:

1. Geo-code the crimes and place them on a map (Figure 10.1).
2. Use the Animal Movement extension from United State Geologic Survey – Alaska to determine the path between robberies. Look for geographic patterns (if any) to help mold your final product (Figure 10.1).
3. Use standard statistical tools to find the most likely date, time, and day of week for the next crime in the series (Figure 10.2).
4. Add the *X* and *Y* coordinates of the robberies to the attribute table and then find the mean and standard deviation of those points (Figure 10.3).

   a. Calculate one Standard Deviation (SD) plus the mean, 2 SD plus the mean, 1 SD minus the mean, and 2 SD minus the mean values. These will identify the corners of the 68 percent probability area and the 95 percent probability area as well as the mean of *X* and *Y* being the center of the crimes.
   b. Map these points and then draw a box from corner to corner to identify the probability areas.
   It is easier to created these as polygon shape files than a graphic.

5. Find the average distance and standard deviation between each crime. This would be the distance from hit number 4 to hit number 5, and then hit number 5 to hit number 6, etc.

   a. Use the average distance figure (at a minimum) to draw a buffer around the last hit in the series
   b. Also use the entire 95 percent range and create multiple buffers when possible.

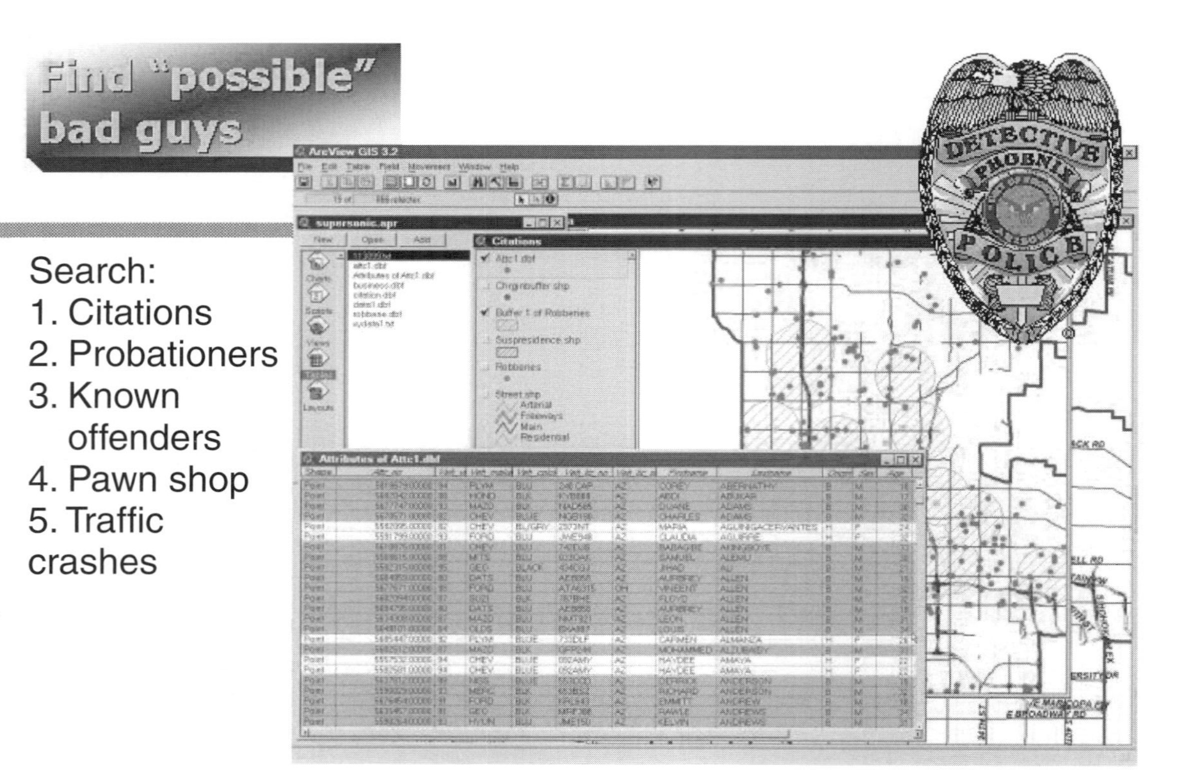

Find "possible" bad guys
Search:
1. Citations
2. Probationers
3. Known offenders
4. Pawn shop
5. Traffic crashes
ArcView GIS 3.2
supersonic.apr
Citations
Attributes of Attc1.dbf
DETECTIVE
PHOENIX
POLICE
Phoenix Police Department –
Research & Analysis Unit

Figure 10.2

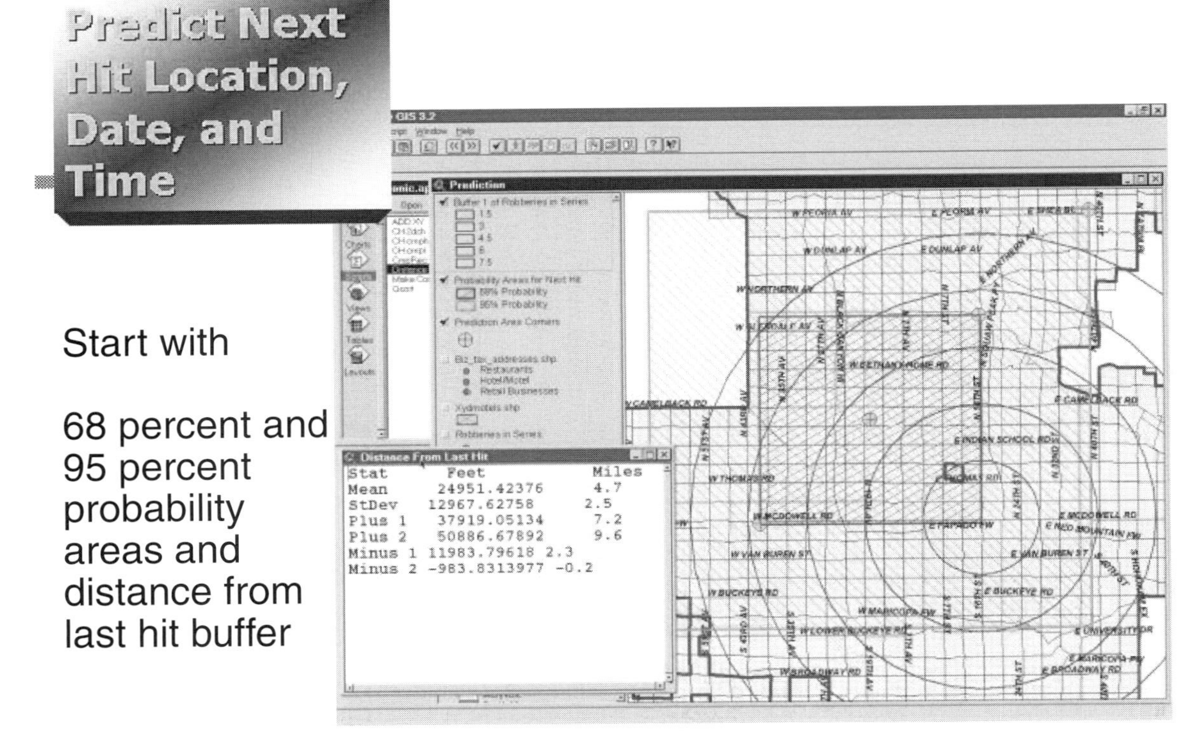
Predict Next Hit Location, Date, and Time
Start with
68 percent and 95 percent probability areas and distance from last hit buffer
Prediction
Distance From Last Hit
Stat Feet Miles
Mean 24951.42376 4.7
StDev 12967.62758 2.5
Plus 1 37919.05134 7.2
Plus 2 50886.67892 9.6
Minus 1 11983.79618 2.3
Minus 2 -983.8313977 -0.2
Phoenix Police Department – Research & Analysis Unit

Figure 10.3

6. Either create a quarter square mile grid theme using the GridMake extension, or use a theme of equal-sized "grids" that may already exist. (Note: Usually a new field is created in the grid layer for each of the scores described below and another field is created that will hold the sum of these scores.)
7. Select the 68 percent area of the probability areas you created in step 4 above. Then use the "Select by theme" tool in ArcView to select the features of the grid layer that intersect the selected attributes of the 68 percent probability area square. Edit the grid attribute table to add the score of 2 to any grids selected, then reverse the selection and score all the other grids with a value of 0.
8. Do the same thing for the 95 percent probability area but use a value of 1 for all the grids selected. One should skip making all other grids equal to 0 to avoid overwriting those you scored as 2 in step 7.
9. Score any grids with a value of 1 that are inside the last hit buffer theme.
10. Any grids that had multiple robberies in the series should be scored with a value of 1 and all others should receive a score of 0.
11. Find the land use theme available for the city in question and use a spatial join between that theme and crimes to determine what kind of land use present in the area the crimes have been committed in before. In my case, I have found that over half have been in the same kind of land use area (high-density commercial, medium-density residential, etc.). Find the most prevalent land use your crimes are in, and then select all the same kinds of land use areas in your land use theme. Use this theme to select the grids that intersect these land use area types and score those grids with a 1 and all others with 0.
12. Use a spatial join between the street theme and the grid theme to get a basic idea of the total length of main streets, arterial roads, freeways, and residential streets in each grid. The street theme and grid theme chosen will determine the best process for this in your jurisdiction. The basic idea here is to score grid with more freeway, main and arterial streets the highest and those with more dead-end residential streets lower because they are less likely as target areas due to ease of escape issues the suspect may consider.
13. Make the grid layer a chloropleth map classified by standard deviation with the grids that scored the highest in the bottom or "hot" color chosen.
14. See if the suspect is hitting the same kind of businesses or in some cases the same store name (i.e. Fry's, Sonic Drive-in, etc.). If the suspect is hitting the same type of business, you can then add a business theme to the project and select only those business types or names that match previous victims AND are in the highest grids (anything between the mean and >3 standard deviations above the mean).

This method appears to be effective for the Phoenix jurisdiction. However, one may need to modify, add, or delete certain parts depending

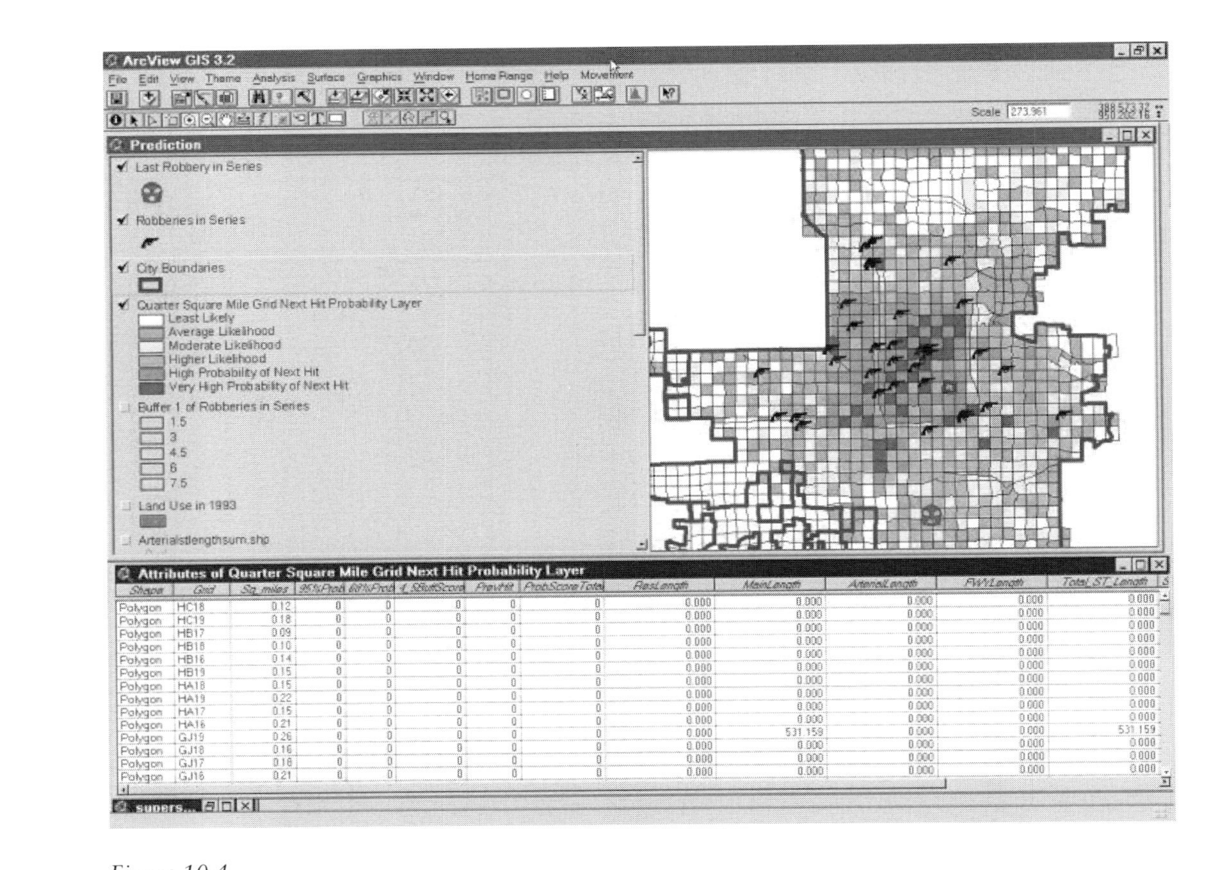
ArcView GIS 3.2
File Edit View Theme Analysis Surface Graphics Window Home Range Help Movement
Scale 273,961
Prediction
Last Robbery in Series
Robberies in Series
City Boundaries
Quarter Square Mile Grid Next Hit Probability Layer
Least Likely
Average Likelihood
Moderate Likelihood
Higher Likelihood
High Probability of Next Hit
Very High Probability of Next Hit
Buffer 1 of Robberies in Series
1.5
3
4.5
6
7.5
Land Use in 1993
Arterialstlengthsum.shp
Attributes of Quarter Square Mile Grid Next Hit Probability Layer
Shape Grid Sq_miles 95%Prob 68%Prob 4_Sqft Score PrevHit ProbScoreTotal ResLength MainLength ArterialLength FWYLength Total_ST_Length

Figure 10.4

Previous Activity Can Help Us Even More:

- Are the targets suspects chose in the past a pattern we can use?

NAME	ADDRESS
GARNETT'S RITE INN	4134 N 7TH AVE
HOLLY INN	2035 N 27TH AVE
PATTY ANNS DRIVE INN	3517 W CAMELBACK RD
SONIC DRIVE IN PHOENIX TH	3314 E THOMAS RD
SONIC DRIVE IN-JESSE OWEN	7440 S JESSE OWENS PKWY
SONIC DRIVE-IN MCDOWELL	742 E MCDOWELL RD
SONIC DRIVE-IN PHOENIX AZ	3310 W BETHANY HOME RD
SONIC DRIVE-IN PHOENIX CA	1537 W CAMELBACK RD
SONIC DRIVE-IN PHOENIX MC	5002 E MCDOWELL RD
SONIC DRIVE-IN PHOENIX 19	9606 N 19TH AVE
SONIC DRIVE-IN PHOENIX 7T	8815 N 7TH ST
SONIC DRIVE-IN PHX AZ MCD	5002 E MCDOWELL RD
SONIC DRIVE-IN 43RD & GLE	4244 N 43RD AVE
SWIZZLE INN	5835 N 16TH ST
TAXI INN	2339 E WASHINGTON ST
330 INN	330 E ROESER RD
ARIZONA SUNBURST INN	6245 N 12TH PL
BLUE PALM INN	4127 N 9TH AVE
DAYS INN I-10 WEST	1550 N 52ND DR
PYRAMID INN	3307 E VAN BUREN ST
RAMADA INN	502 W CAMELBACK RD
AUTO ZONE #2732	2601 E MCDOWELL RD
AUTO ZONE #2752	2418 E SOUTHERN AVE
AUTO ZONE #2764	3101 N 19TH AVE
AUTO ZONE #2778	3908 E THOMAS RD
AUTO ZONE #2789	8515 N 7TH ST
AUTO ZONE #2790	3359 W VAN BUREN ST
AUTO ZONE #2792	2705 W CAMELBACK RD
AUTO ZONE #2793	6725 N 35TH AVE
AUTO ZONE #2795	4319 W INDIAN SCHOOL RD
AUTOZONE #2725	221 E CAMELBACK RD
AUTOZONE #2762	4832 S CENTRAL AVE

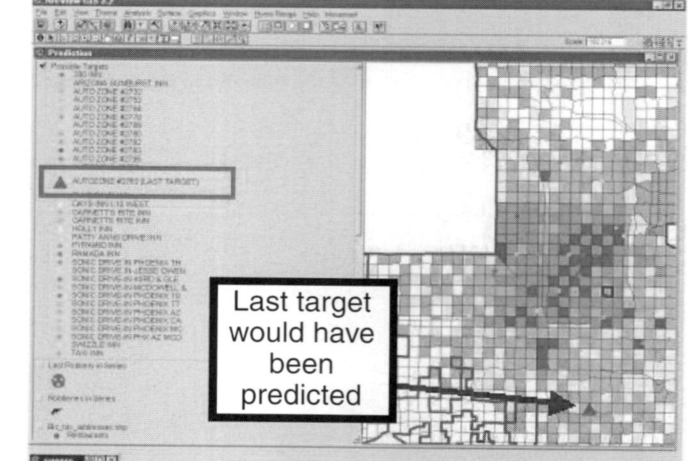

Phoenix Police Department – Research & Analysis Unit

Figure 10.5

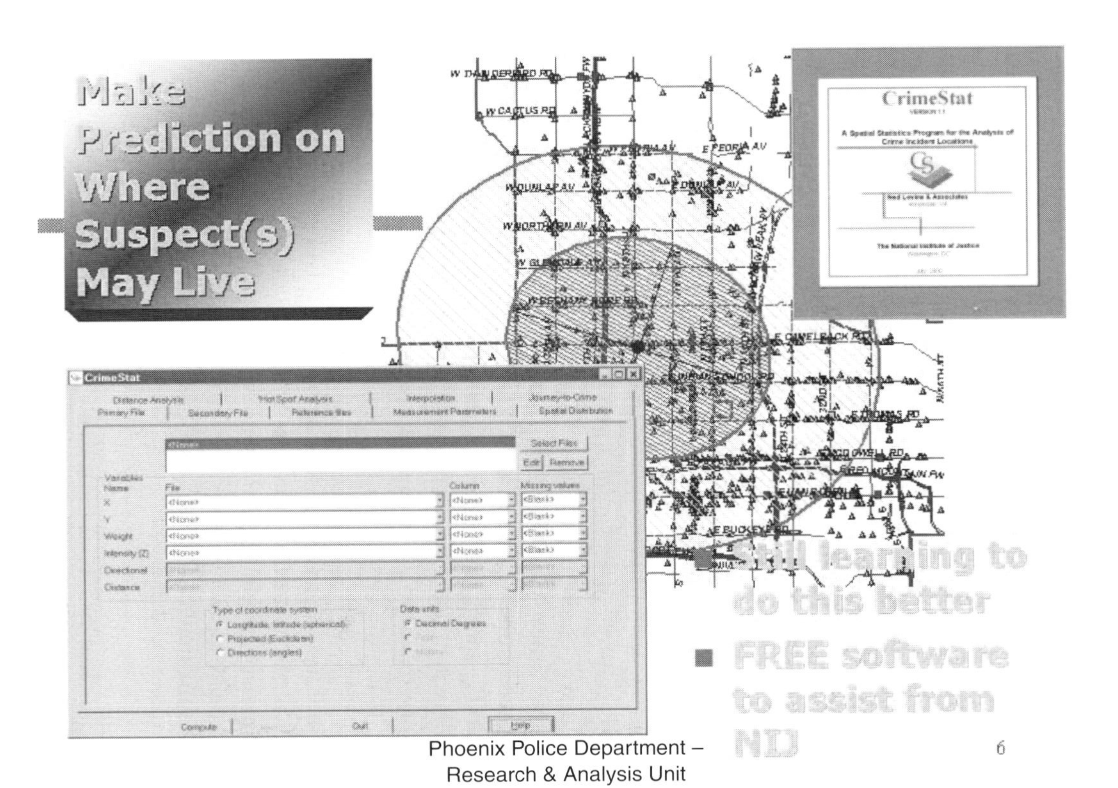
Make Prediction on Where Suspect(s) May Live
CrimeStat
A Spatial Statistics Program for the Analysis of Crime Incident Locations
Ned Levine & Associates
The National Institute of Justice
Still learning to do this better
FREE software to assist from NIJ
Phoenix Police Department – Research & Analysis Unit
6

Figure 10.6

on the value they provide one's jurisdiction (see Figures 10.4–10.6). This process will also be modified for other crime types in the future and may not be applicable to all crime types in Phoenix.

What we are doing

The basic focus for our unit is to try and take the massive amounts of data more available to users within the police department. We are committed to making GIS the conduit for transfer of the data to the users. There are many obstacles in the way, such as technology problems, large data sets and in some cases, learning curves for new software and products. We are implementing a process to enter robbery data into the ATAC program for trend analysis in robbery offenses so that perhaps we can find a robbery series before the Robbery Detail brings it to our attention. This will allow us to take a more proactive approach to crime analysis in this area.

We currently provide a Crime View class every month and will soon be teaching an Introduction to ArcView GIS class on a monthly basis. We also plan on trying to get AZPost certification of the CrimeView and ArcView teaching curriculum and work on a crime analysis and data quality class for our regular academy and in-service training functions. We are implementing GIS into every detail and unit in the Phoenix Police Department and believe that it will be the backbone of our crime analysis and community-based policing processes within the next three years. Crime analysis involves five components. They are COLLECT, COLLATE, ANALYZE, DISSEMINATE, and EVALUATE. We believe that crime analysis and community-based policing are one in the same process for improving the quality of life. Adding GIS technology to our efforts can only enhance, strengthen, and add quality to the very fiber of the Phoenix Police Department. Methods in policing are changing, and although there appears to be no lack of job security for police departments, GIS will be in the forefront of policing activities. It assists us in doing our jobs: fighting crime, arresting criminals, helping neighborhoods restore their community pride, and helping themselves. GIS provides accurate and timely data to ensure good decision making at all levels of police administration.

Editors' note

Although the Phoenix Police Department choose to geo-code business information, already mapped datasets showing various types of business by standard industrial classification and containing business names are available from vendors such as Claritas. Not all US cities have available data and not all types of businesses are represented, but such robbery targets as fast food franchises and gas stations have been included in this data set, which is of course not available free of charge.

11 Crime mapping at the Boston Police Department

Carl Walter

Boston, Massachusetts is a sophisticated user of GIS technology throughout its municipal government. This sophistication has developed over many years in what is arguably the most cultured and highly educated major city in America. It is fitting that the home of great universities such as Harvard and Boston University is also a great example of application of GIS technology in law enforcement.

(Editor)

Over the past several years, crime mapping has emerged as an important tool for supporting problem solving and prevention efforts at the Boston, Massachusetts Police Department. The Boston Police Department has over 2,000 sworn officers serving a city with a population of 574,283. Through the use of the GIS technology, the Department has made great strides in advancing crime mapping and applying it to support their mission of *Neighborhood Policing*. Today, the Department has a reputation as one of the most advanced law enforcement agencies in the nation in its use of GIS.

During the year 2000, representatives from the Department's Office of Research and Evaluation worked on the development and testing of crime-mapping applications for the US Department of Justice, provided expert courtroom testimony using GIS technology, and presented papers at various crime-mapping training seminars and conferences throughout the country.

However the most notable advancement during 2000 was the development of CrimeShow – the Department's customized crime-mapping application. The Boston Police Department developed CrimeShow in partnership with ESRI – Boston. Created in MapObjects, this application was designed to fulfill dual needs: state of the art presentation capabilities along with a complex and robust analytical component. The CrimeShow application is used throughout the department and has been integrated as the primary presentation tool for use during the Department's bi-monthly Crime Analysis Meeting (CAM). Figure 11.1 provides a flow chart and discussion of the CAM process. CrimeShow includes basic mapping and charting functionality, as well as links to on-line incident reports, booking sheets, mug shots,

Boston Police

Department

Problem Solving

Best Practices

Accountability

THOMAS M. MENINO
Mayor of Boston

PAUL F. EVANS
Police Commissioner

THE CAM PURPOSE

To provide a useful forum which facilitates informed decision-making through the use of widely shared data, crime analysis techniques, and innovative problem-solving strategies supported by Geographic Information System (GIS) technology.

THE CAM OBJECTIVES

1. ***Identify*** problems using appropriate data and GIS-based crime analysis techniques;
2. ***Support*** ongoing creative problem-solving and crime prevention efforts;
3. ***Develop*** a shared pool of knowledge, information, and "best practices";
4. ***Enhance*** the awareness of the problems that are identified and solved successfully at the District level; and
5. ***Share*** the results of these anti-crime strategies and innovative solutions throughout the Department.

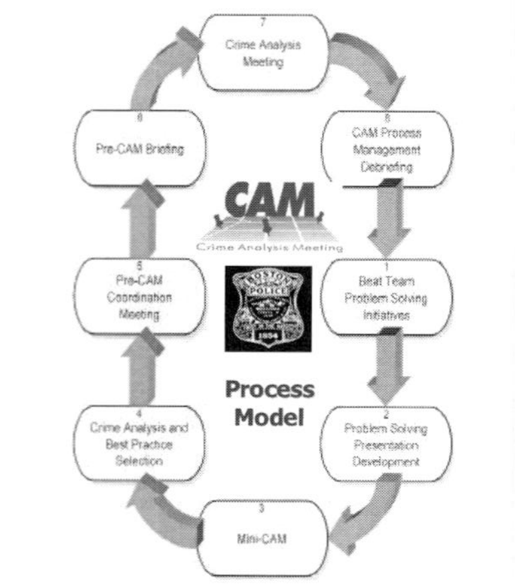

THE CAM PROCESS

1. **Beat Team Problem Solving Initiatives**
   Beat Teams across the City develop innovative strategies to address problems on their beat – using the SARA model (Scanning, Analysis, Response and Assessment).
2. **Problem Solving Presentation Development**
   Once a problem has been addressed through the SARA process, Beat Teams prepare a presentation that provides an overview of their strategy that is shared at a District mini-CAM.
3. **Mini-CAM**
   Mini-CAMs are used at the District level to identify problems through GIS-based crime analysis, and develop and share best practices across the District's Beat Teams.
4. **Crime Analysis and Best Practice Selection**
   After each Mini-CAM, the District Captain and his/her senior personnel review the best practices to select the one most suitable for sharing with Department personnel from across the city. That Beat Team is tasked with giving their presentation as a part of their District's presentation at the next CAM.
5. **Pre-CAM Coordination Meeting**
   Prior to each CAM, District personnel meet with the Office of Research and Evaluation to finalize the District's crime analysis and best practice presentations and develop a focused agenda.
6. **Pre-CAM Briefing**
   Prior to each CAM, the Office of Research and Evaluation briefs the Command Staff on emerging crime trends, issues for follow-up from the presenting District's prior CAM, and the upcoming CAM's agenda.
7. **Crime Analysis Meeting**
   At each CAM, Districts provide an overview of emerging crime trends and what actions are being taken to address them, and present on a successful beat-team strategy selected from the District's Mini-CAM.
8. **CAM Process Management Debriefing**
   The CAM Process Management Team meets monthly to address needs for improvement to the CAM process and identify best practices warranting greater distribution (e.g., through newsletters, videos etc.)

Figure 11.1

and 360 degree interactive photographs (all of which are activated by selecting points on a map).

CrimeShow, and other GIS technology, have proven to be invaluable in the support of problem-solving efforts and informed decision-making by police officers and managers throughout the Boston Police Department. The following materials illustrate a CrimeShow presentation of the type that would be used in a departmental crime analysis meeting (Figures 11.2–11.11).

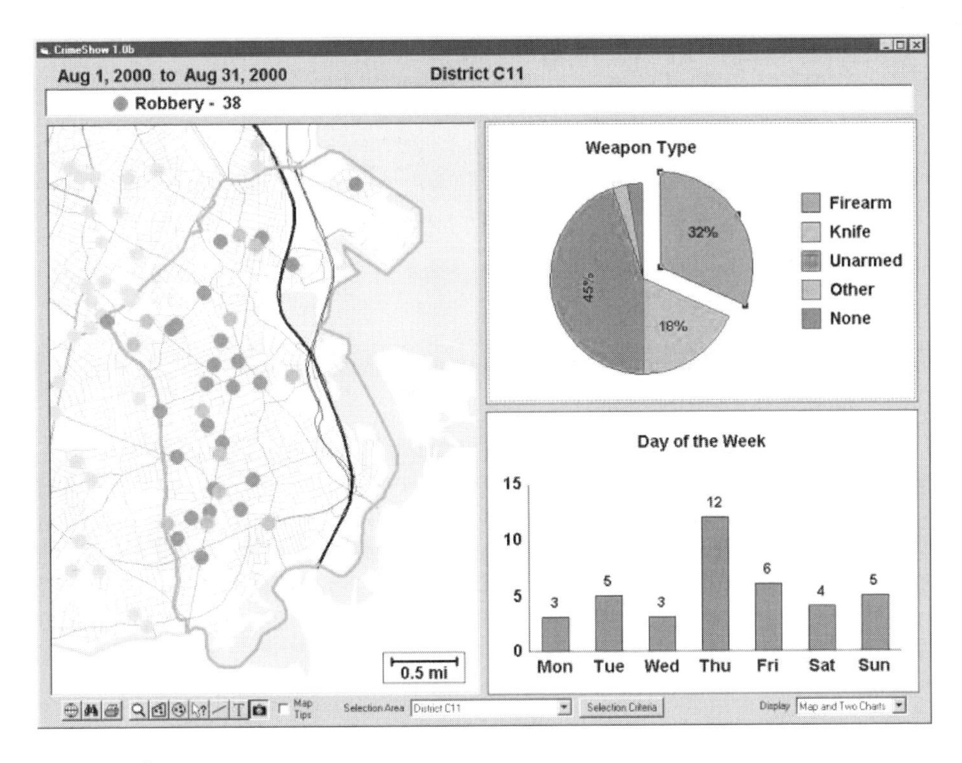

Figure 11.2

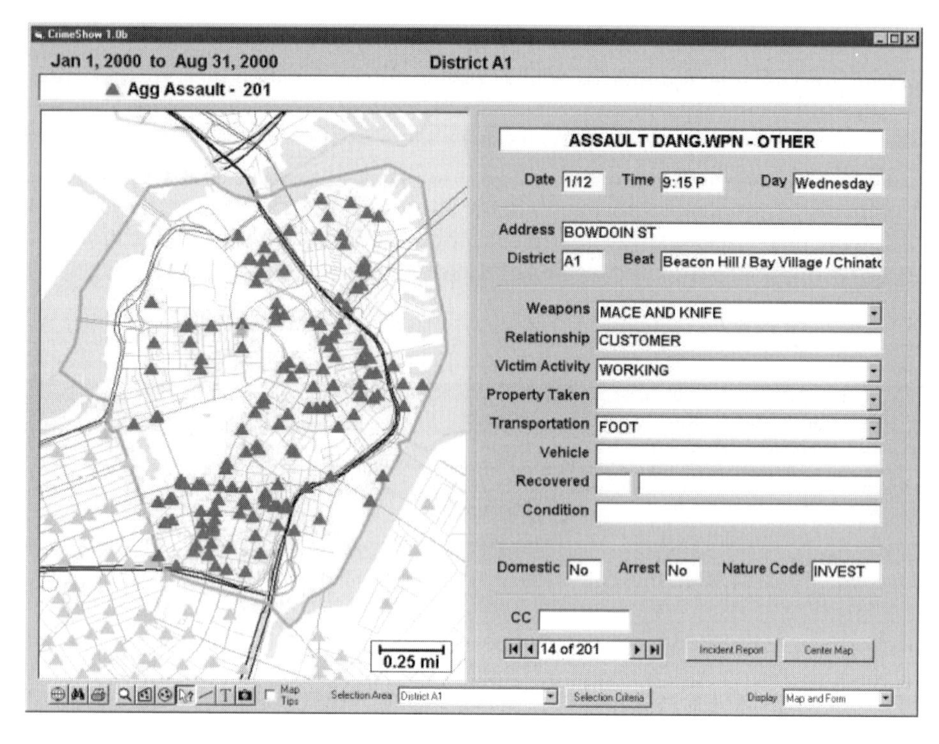

Figure 11.3

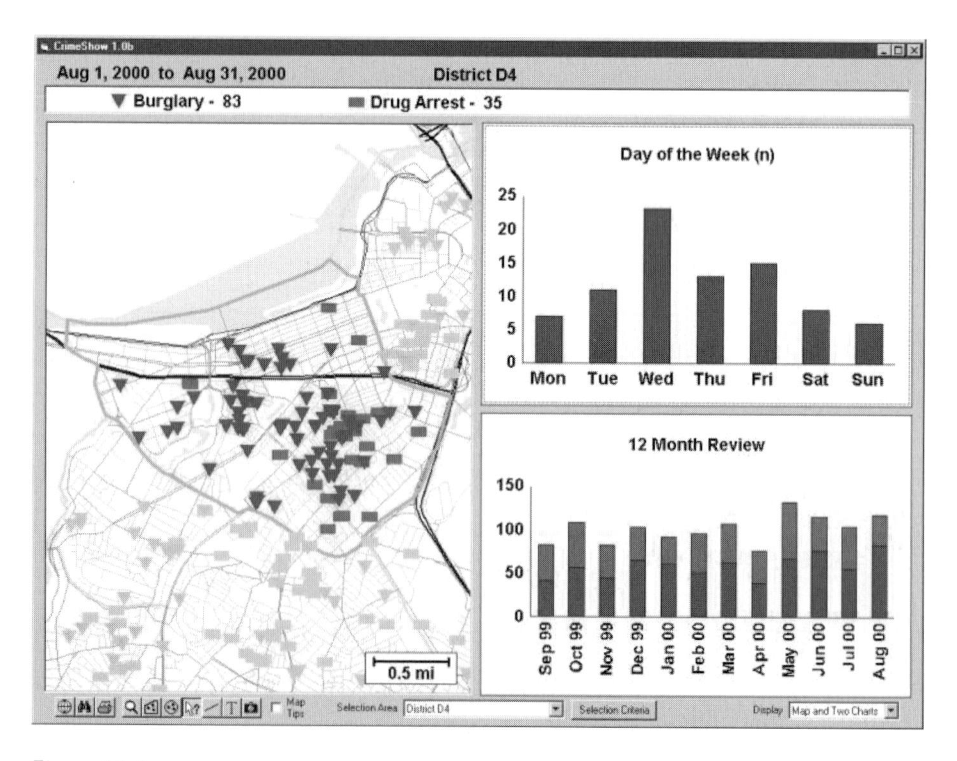

Figure 11.4

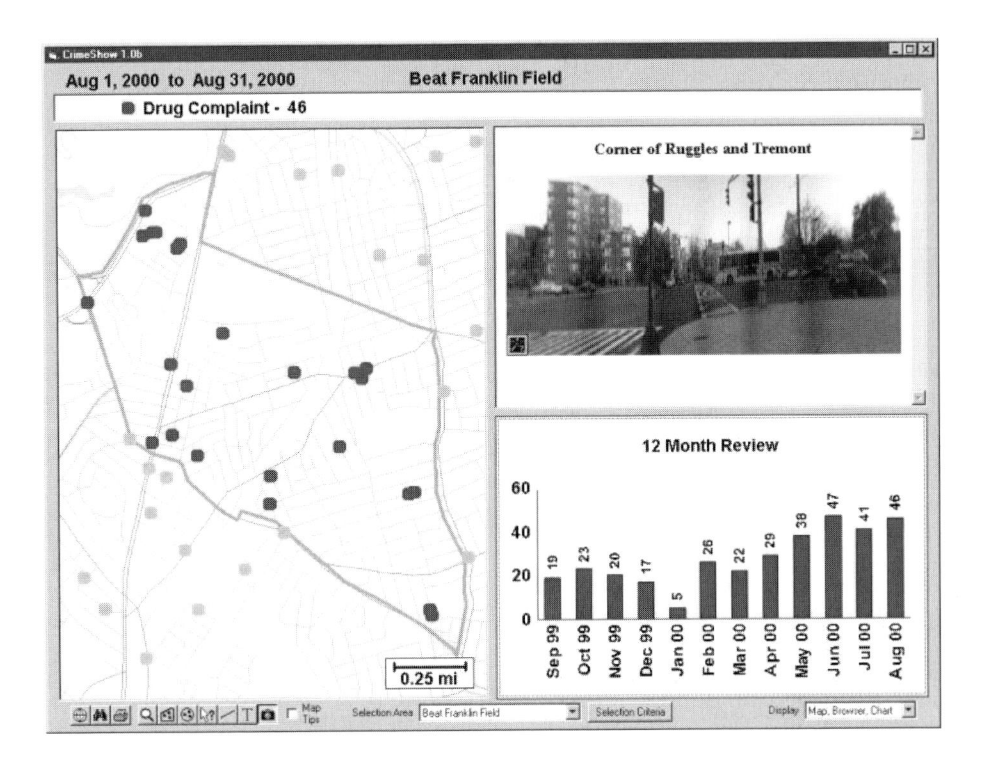

Figure 11.5

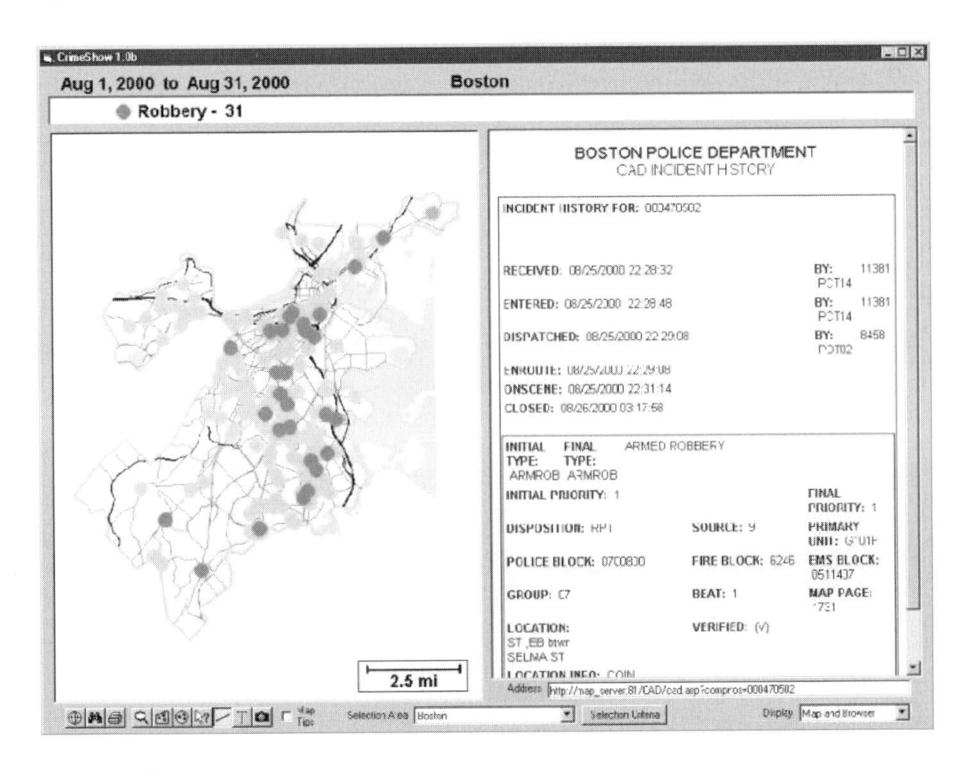

Figure 11.6

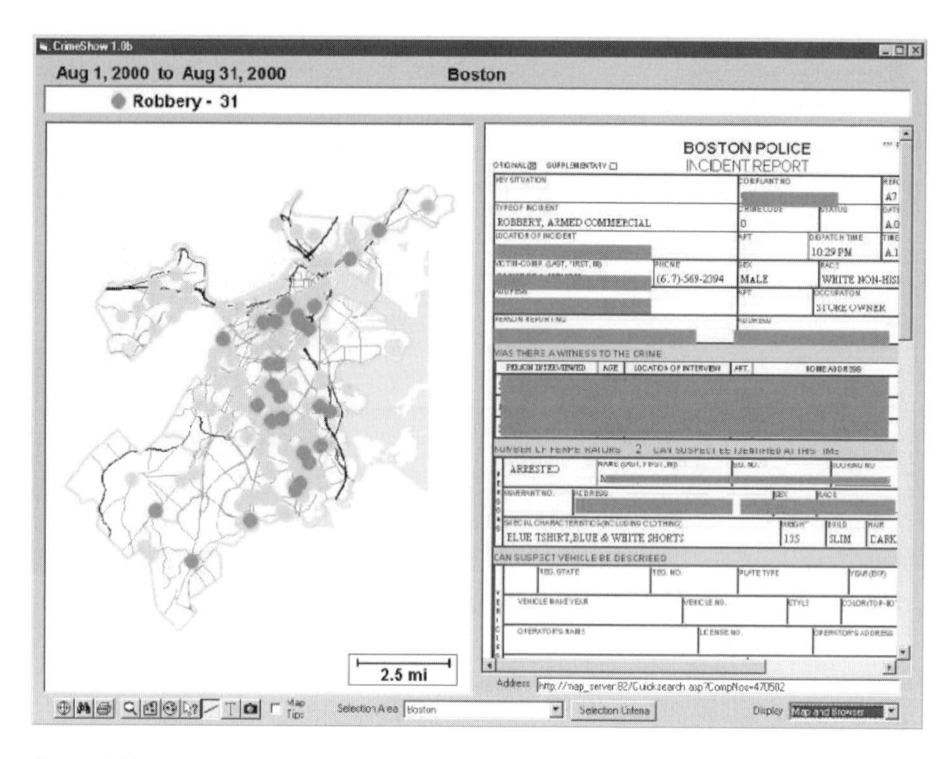

Figure 11.7

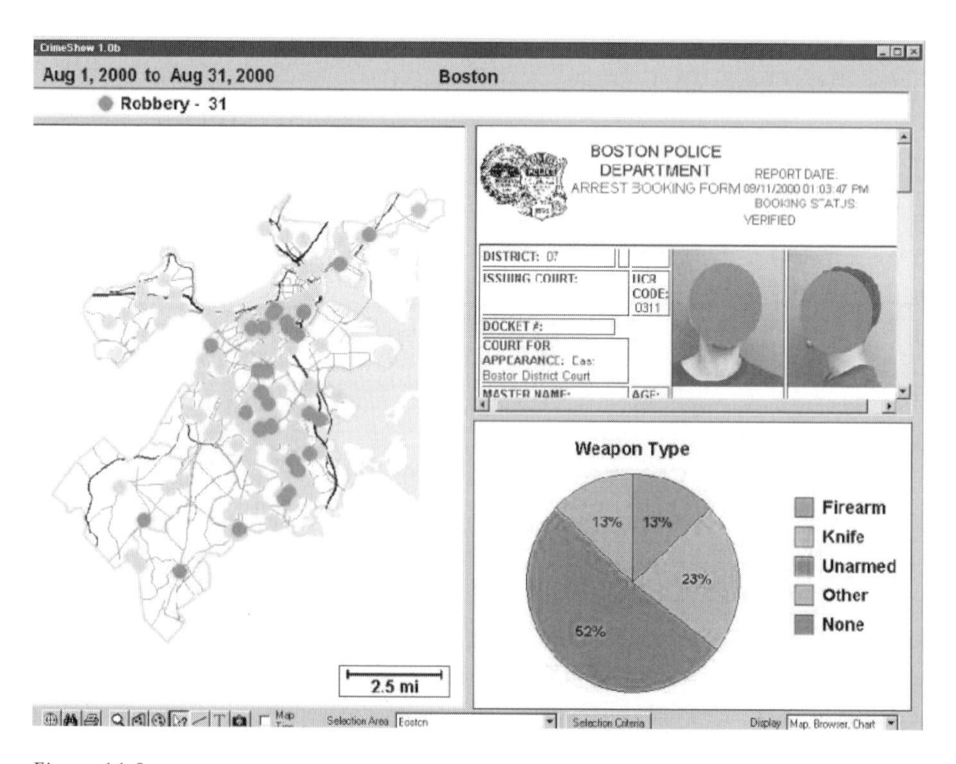

Figure 11.8

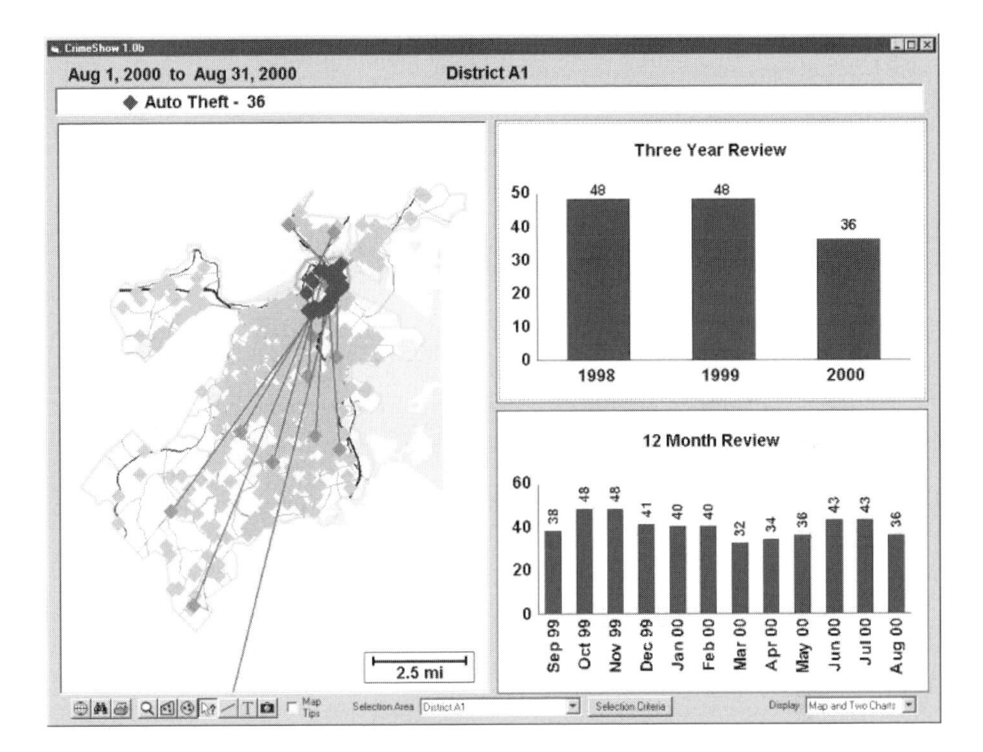

Figure 11.9

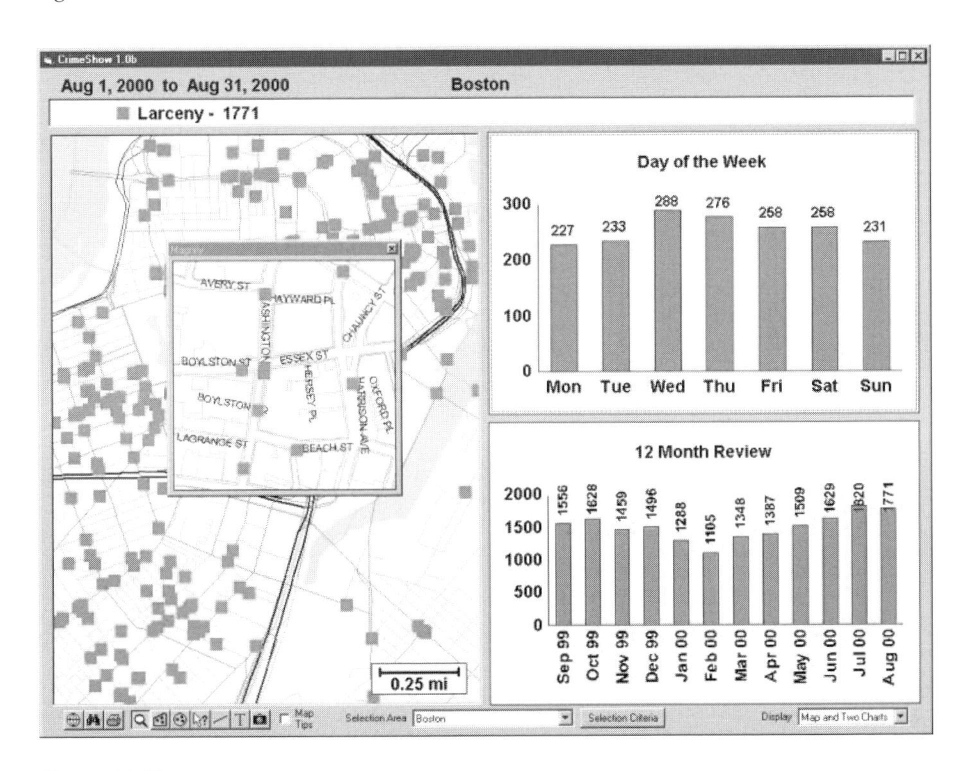

Figure 11.10

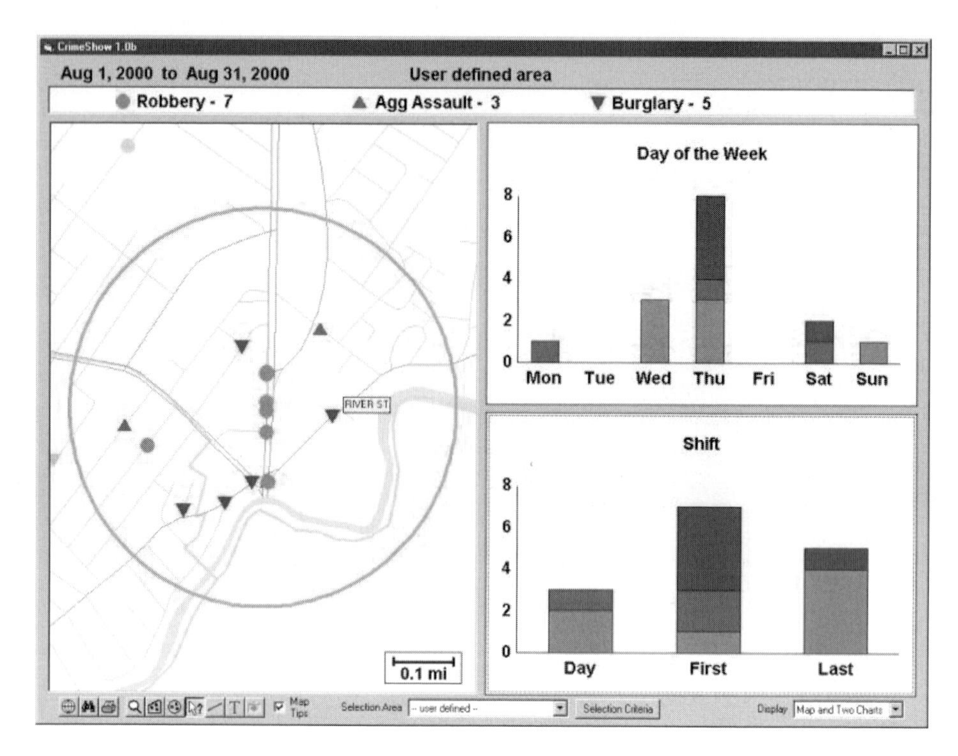
CrimeShow 1.0b
Aug 1, 2000 to Aug 31, 2000
User defined area
Robbery - 7
Agg Assault - 3
Burglary - 5
RIVER ST
0.1 mi
Day of the Week
8
6
4
2
0
Mon
Tue
Wed
Thu
Fri
Sat
Sun
Shift
8
6
4
2
0
Day
First
Last
Map Tips
Selection Area
-- user defined --
Selection Criteria
Display
Map and Two Charts

Figure 11.11

12 Apprehending murderers in Spokane, Washington using GIS and GPS

Mark Leipnik, John Bottelli, Ian Von Essen, Ariane Schmidt, Laurie Anderson, and Tony Cooper

The two cases discussed in this chapter illustrate some of the most exciting applications of geo-spatial technologies in law enforcement. In particular, the increasingly important role that GPS will play in the future of law enforcement is foreshadowed in the story that is unfolded below.

(Editor)

Introduction

This chapter describes two distinct cases where geo-spatial technologies were applied to the investigation of homicides in the metropolitan area of Spokane, a city in the Eastern portion of Washington State. In one case (*The Brad Jackson Case*), in which GPS and GIS played a pivotal role, a man suspected in his wife's earlier disappearance and the subsequent disappearance of his nine-year-old daughter under suspicious circumstances was ultimately arrested and convicted of murder based on GPS data, processed using GIS. The GPS data ultimately led investigators to both a temporary grave five miles from Brad Jackson's Spokane Valley home and a more isolated permanent grave in a remote area of southwestern Stevens County, Washington about sixty miles away. This second grave contained the suffocated body of his young daughter. He had unknowingly led investigators to the location where he had exhumed his already dead daughter and then transported her to what he must have felt was a safer hiding place where he reburied her. Jackson successfully "stonewalled" detectives in his wife's, technically still unsolved disappearance. It seems probable that he would have evaded prosecution in his daughter's disappearance as well, without evidence that depended directly on geo-spatial technologies. Spokane County GIS specialists used GPS data coupled with GIS base-maps to map out routes and waypoints from Jackson's secretive journeys and extract accurate coordinates of gravesites. It can be convincingly argued that only by using these geo-spatial technologies, could Spokane County Sheriff's Department investigators have obtained the evidence that led to the arrest and conviction of Brad Jackson.

In the other case, both GIS and GPS also figured prominently, although with the unusual twist that the suspect was using the GPS technology, possibly to facilitate his own nefarious activities. In this unique case (*The Robert Yates Case*) GIS was used to map locations in the Spokane region where the bodies of ten murdered prostitutes were discarded. In this case, the plastic bags that were used to cover the victims' heads provided an important spatial clue since they were imprinted with the supermarket's name and therefore the spatial distribution of these supermarkets also became a possible clue to the "home range" of the killer. To study the distribution of both "body dump" locations and locations of supermarkets from which plastic bags found on the bodies of victims were possibly obtained, a spatial modeling approach was employed by analysts. This approach utilized existing geo-statistical models designed to predict the "home range" of predatory wildlife to try to predict the home range of this human predator. Although this spatial analysis did not "solve" the case, the predicted "home range" of the killer was in the same neighborhood of Spokane in which the killer actually resided, and the analysis was helpful in narrowing the focus of the investigation.

In a fascinating twist, the confessed serial murderer himself possessed and actively utilized a GPS unit. Seventy-two waypoints stored on this unit were actively investigated to determine if the killer used the unit to mark body dump locations in isolated areas of Washington and Oregon. The dates and routes of journeys recorded by the GPS unit may provide investigators clues into other unsolved homicides, and in any case helped to establish the movements of the killer. These spatial clues are being investigated using the coordinates of the extracted waypoints and GIS-generated maps. Possible links have been established between Yates and several unsolved homicides in other parts of the Pacific Northwest. At any rate, the use of GPS technology by a vicious and long-active serial killer should send a wake-up call to law enforcement that geo-spatial technologies are also being adopted by perpetrators of serious crimes, perhaps even before they are being employed by many law enforcement agencies.

The Brad Jackson case

Background

On October 26, 1999, Spokane County Superior Court Judge Kathleen O'Connor granted Spokane County Sheriff's Department detectives the authority to magnetically attach a Silent Position Monitor 2000 GPS unit to the underside of Brad Jackson's 1995 Ford pick-up truck as well as his 1985 Honda Accord sedan for a ten-day period. On November 2, 1999 Judge O'Connor granted a ten-day extension of her order. These devices were intended to help track the movements of these vehicles. She issued these authorizations under the same legal doctrine that allows use of surveillance cameras and wiretaps of phone conversations or emplacement and retrieval of electronic eavesdropping devices, including those placed in automobiles. For such authority to be granted a judge must be convinced that "probable cause" exists.

The reason that probable cause existed in this case, was that not only had Brad Jackson reported the disappearance of his 9-year-old daughter Valiree from his parents' home under suspicious circumstances on October 18, 1999, but seven years earlier his wife Roseann had disappeared without a trace as well. Law enforcement personnel and community members initiated a sweeping search for Valiree but no sign of her was found. Spokane County Sheriff's Department investigators were suspicious that foul play had occurred in the earlier case, but Jackson, who was the last known person to see his wife alive, was unshakable in his denials of culpability for Roseann's as yet unsolved disappearance. Discrepancies in his statements about Valiree, coupled with the earlier disappearance of his wife made Brad Jackson the prime suspect in this new case and provided the probable cause for the attachment of the GPS tracking units to his vehicles.

After eighteen days the units were retrieved by authorities and with the assistance of GIS specialists and the GIS network manager for Spokane County, the movements of Brad Jackson's vehicles were reconstructed and the locations of the gravesites ultimately found. Thus, the laborious and often fruitless effort to "shadow" suspects like Brad Jackson over a prolonged period was obviated. Even had Spokane County Sheriff's Department investigators had the resources to trail Jackson, and the skill to avoid arousing his suspicions, it is highly unlikely that they could have followed him into the remote areas where he had buried his victim (in all likelihood his *second* victim).

GPS data management

Spokane County GIS Specialist, Tony Cooper, using data provided by Don McCabe of the Spokane County Sheriff's Department, completed the initial conversion and mapping of the GPS data generated from the Brad Jackson case. The GPS unit chosen was made by Integrated Systems Research of Baltimore, Maryland and was a silent position monitor model SPM2000. The chosen unit had a wireless modem built into it, and it was planned that data be downloaded once a day, but telemetry proved unreliable. Data was retrieved from the unit by physically removing the unit from the vehicle. The GPS, unit attached magnetically to the undercarriage of Jackson's vehicle, was programmed to record locations every twenty seconds, which caused points indicating tracks to be spread further apart during high-speed travel. However, it was tracking of points generated by low-speed travel along logging roads that provided the most important clues in the case. Of the day-to-day travel recorded on the GPS units, two sets of points related to two journeys stood out as suspicious. One set of points showed the route of Jackson's pick-up truck from his home in the Spokane Valley to his brother's house in Pend Oreille, County, and then from there to a remote location on a logging road in southwestern Stevens County. These points were associated with a journey that occurred on November 6, 1999. A second set of points was generated by a journey that occurred on November 10, 1999. The second set of points when mapped on an appropriate GIS base-map, indicated a journey that led first to a somewhat isolated location in the Spokane Valley about five miles from the suspect's home and then went directly back to the same remote location in Stevens County. This led investigators to postulate that Jackson's first trip was designed to scope out an isolated spot to bury (actually rebury) his daughter's body, while his second journey involved disinterring the body from a temporary grave in the Spokane Valley and then traveling laden with the body to a remote final gravesite in Stevens County. See Figures 12.1 and 12.2 for maps of Brad Jackson's travels as generated from data retrieved from the GPS unit attached to the suspect's pick-up truck.

GIS methodology

GIS specialist Tony Cooper, under the guidance of GIS Network Manager Ian Von Essen, both with the Spokane County Government used Arc/Info Version 7.1 software from ESRI. This software was used to help process GPS points and to generate base-maps to help track the travels of Jackson's pick-up truck in a comprehensible spatial context. Arc/Info was also used by Spokane County GIS specialist John Bottelli to generate all the final revised versions of the GIS maps that were to become crucial exhibits in Brad Jackson's trial.

The GPS points related to the travels of Jackson's truck (and thus the prosecutors could persuasively argue Jackson's *own* travels) were generated into a map layer using the Arc/Info command "*Generate.*" Then the Arc/Info command "*Joinitem*" was used to join GPS attribute data (time/speed fields) to the associated *x, y* coordinate points. Once this process was completed, the plotting of the *x, y* coordinate points led outside of Spokane County into forested rural areas of Pend Oreille and Stevens Counties which lie to the north of Spokane County. Because of this fact, digital GIS data sets of roads and hydrology (lakes and streams) had to be acquired for Pend Oreille and Stevens County before the maps could be generated. Once the data was acquired, maps of the two sets of GPS data were then generated and used by detectives for field investigations and in the subsequent trial.

Status

Based largely on the evidence provided by geo-spatial technologies, investigators discovered the carefully hidden body of Jackson's daughter and Brad Jackson was arrested and charged with 1st degree murder. On October 5,

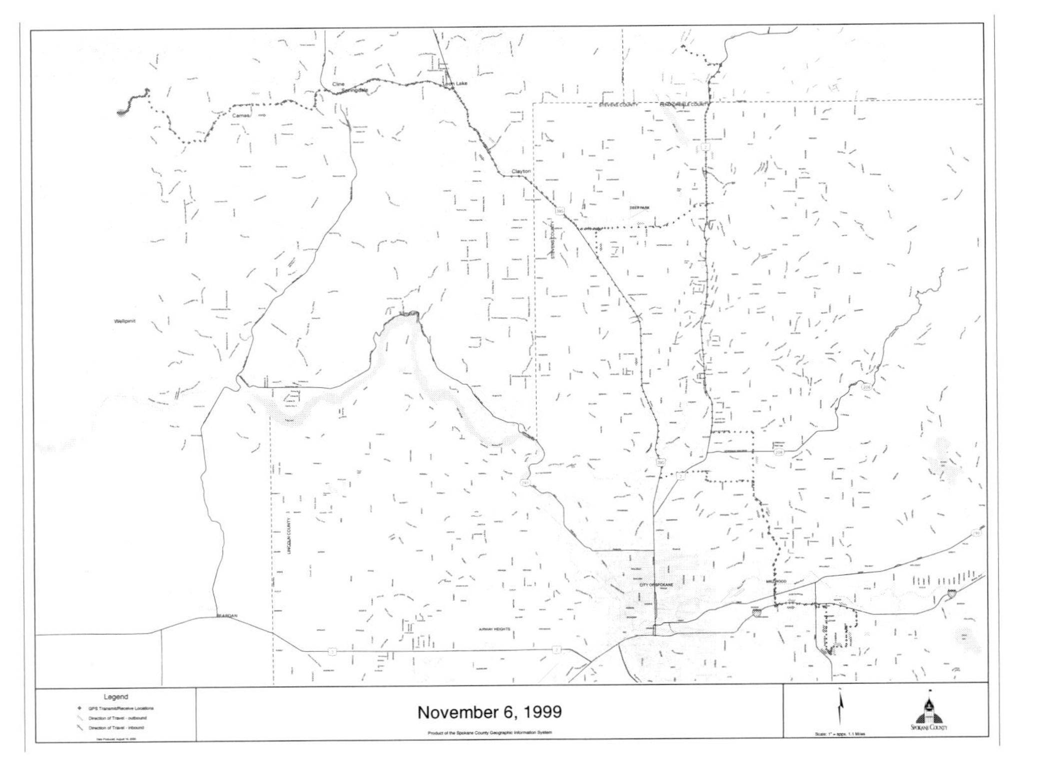

Figure 12.1 This map of the Spokane region and surrounding counties shows the track of murder suspect Brad Jackson's pick-up truck. The truck was fitted with a GPS-based tracking device. This trip was made on November 6, 1998. The trip involved a visit to Jackson's brothers house in Pend Oreille County about 60 miles north of Spokane, then terminated at a spot along an isolated logging road in Stevens County, Washington, prior to a return journey directly to Spokane.

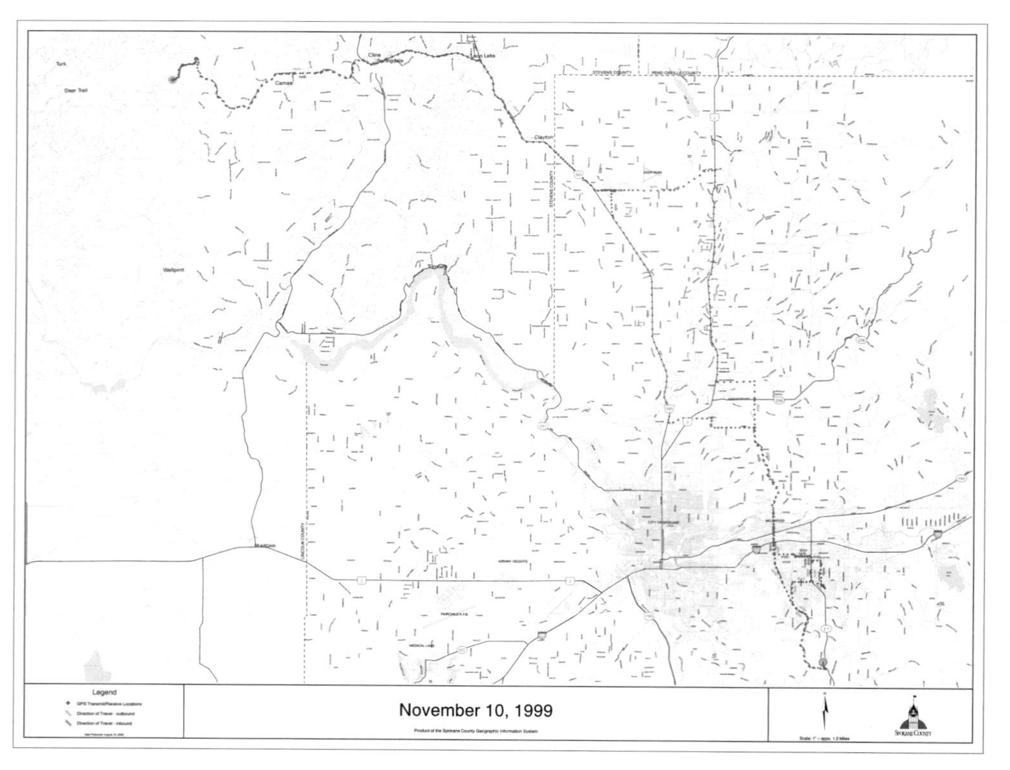

Figure 12.2 This map portrays the travels of Brad Jackson's pick-up truck on November 10, 1998. He first traveled to a somewhat isolated spot five miles from his home in Spokane, then after an interval the truck headed directly to the same spot on the isolated logging road visited on 6th November. After downloading these points and mapping them investigators were guided to a shallow empty grave at the site near Spokane and a grave containing the body of Jackson's missing nine-year-old daughter.

2000 a Spokane County Jury convicted him and on November 9, 2000 he was sentenced to fifty-six years in prison.

Although the evidence produced by the use of GPS to track Brad Jackson seems damning, Jackson's defense attorney David Hearrean has been quoted by the local Spokane Paper as challenging the validity of use of GPS to "spy" on suspects and may appeal Jackson's murder conviction in part on those grounds. Mr. Hearrean states: "This is a fairly new legal issue in our state that needs to be tested in our courts." Mr. Hearrean contends that use of GPS to track suspects without their knowledge might involve an invasion of privacy rights and might not meet the legal test of finding the "least obtrusive means" for police to gather information about suspects. Although, one could argue that GPS is far less obtrusive than using one or more police officers to shadow a suspect. In fact, because of its silence, small size and unattended operation, GPS units can be used in an unobserved manner and hence they are capable of being more effective than traditional methods in cases like Jackson's.

Other cases testing the legality of GPS use are making their way through the US courts. Cases in recent years involving a Miami, Florida cocaine smuggler and a serial arsonist in Kansas City, Missouri have been solved using hidden GPS units. In a Montana case, where a GPS unit was attached to the underside of a suspected marijuana grower's vehicle to track his movements to isolated marijuana groves in a National Forest, a Federal Appeals Court has ruled that investigators need not even obtain a warrant beforehand. The 1999 ruling of the appeals court found that there was no reasonable expectation of a "right of privacy" for the undercarriage of a car. Thus, we can expect to see more cases where GPS is used to track the movements of criminals, although few will have as definitive a denouncement as the Brad Jackson case.

The Robert Yates case

Background

The Robert Yates case is too involved, not to say gruesome, for extended recapitulation in a technical book of this kind. Look to authors of "true crime" exposes to feature this case of an organized serial killer in all its deadly detail. In summary, Robert Yates, an Army National Guard helicopter pilot and metalworker living in Spokane, and a middle-class husband and father of four led a second secret life of depravity. He preyed principally on prostitutes, and has confessed to ten murders in Spokane while evidence links him to an additional murder in Spokane and two earlier murders in Walla Walla, Washington. These thirteen murders do not include two murders in southern Washington State in which he has recently been charged. Furthermore, he is suspected in three earlier murders in Germany and a murder in Alabama. However, since his earlier crimes did not fit a recognizable pattern they were not attributable to a single individual until after Yates' arrest.

Beginning in 1996 however, Robert Yates settled into a distinctive *modus operandi*. He cruised the streets of Spokane in a white corvette and solicited streetwalkers and call girls who were subsequently killed. His trademark, after the first rather bloody murder in Spokane, was to place one or more grocery store plastic shopping bags over the heads of his victims prior to disposal of the bodies. Such a distinctive feature along with Ballistics evidence helped investigators to discern that a serial killer was at work, rather than the deaths being part of a random toll of killings of practitioners of an ancient and risky profession. Perhaps the reason that plastic bags were employed was to avoid contamination of Yates' car with blood or other potential sources of DNA evidence. At any rate, as the number of murdered prostitutes found in the Spokane region who fit this pattern began to grow, a Homicide Task Force was organized jointly between the Spokane Police Department and the Spokane County Sheriff's Office. As the death toll mounted, investigators pursued many avenues. For example thousands of interviews were conducted and the results and tips were computerized and cross-indexed. Female decoy officers were placed in various "red light" areas in the city and assistance was sought from the public, the media, and "working girls." The police were on the lookout for suspects arrested with plastic bags and decoy officers scanned "cruising" vehicles for plastic bags as well.

Spatial analysis approach

Based on a suggestion from Spokane County Sheriff's department detective and Homicide Task Force member Fred Ruetsch, an effort to use the technique of geographic profiling was made. Police planners Ariane Schmidt and Laurie Anderson worked on putting available data together, and assistance was also sought from Kim Rossmo at the time affiliated with Simon Frasier University and the Vancouver Police Department. Dr Rossmo, has developed the reputation as one of the world's outstanding experts on geographic profiling – a technique since adopted in several serial murder investigations and used by the FBI behavioral sciences office, among other law enforcement agencies. Unfortunately, the circumstances of the Yates case were not ideal for use of Dr Rossmo's technique. Nevertheless, investigators decided to use GIS to attempt to identify spatial patterns in the case in an effort to focus the investigation or perhaps better deploy undercover operatives.

Because of the convoluted nature of the liaisons between prostitutes and their clients, and a certain lack of cooperation from active prostitutes, police were not able to track the movements of the murdered women or even get a good understanding on the areas they frequented. There were, however, some other locations that could be more readily determined. Firstly, the locations where the bodies of the murdered women were dumped could be determined accurately as well as the time that they died and the time they were discovered. Secondly, the killer inadvertently left another valuable clue.

Starting in December 1997 with Shawn Johnson, bags covered the heads of six subsequent victims including the most recent victim Connie Ellis, found October 14, 1998. Two shopping bags were found in the grave of Darla Scott, believed murdered in November 1997. The killer initially used three bags to cover the victim's head, and later used only two. In each case, where a shopping bag (or several bags) were used to cover the victim's head, the grocery store that provided the bag could be determined. It turns out that in the course of ten murders the killer used shopping bags from five separate supermarkets in Spokane. These stores were scattered across Spokane. In some cases multiple units of a given chain existed; in other cases, only a single store (the Super One grocery store) was present in the Spokane area. There was a cluster of all five stores located on Spokane's South Hill neighborhood. Merely mapping the location of stores and body disposal locations gave investigators some insights into the possible areas in which the killer was active. In fact, one might suppose that body dump locations would tend to be in less frequently traveled areas, near where the prostitutes were accosted. One might also suppose that the supermarket locations might point to an area in proximity to the residence of the killer. For, one would suspect that the bags were obtained in the course of routine shopping trips by the killer (or perhaps in retrospect by his wife who may have done the bulk of the family shopping). Now many studies of consumer behavior would tend to support the idea that ordinary people are more likely to shop at markets nearby their homes and/or on the route from work to home, than at distant stores. See Figure 12.3 for GIS-generated map of Spokane Washington with body disposal locations and supermarkets where bags found with victims could have been obtained.

The challenge for task force members, including the GIS specialists with the County government assisting the task force, was how to objectively analyze this intriguing data so as to help point out the "home range" of the killer, either in his predatory mode or as a ordinary grocery shopper. To undertake this analysis, the GIS analysts had an inspired idea. Rather than trying to misapply existing crime analysis software to a problem its originators indicated was inappropriate, they decided to use an existing GIS-based geo-statistical model compatible with Arc/Info. This model was developed by the US Department of the Interior in Alaska. What a remarkable insight to apply a model designed to determine the home range of animals (including predatory ones) to a predatory human.

The ArcView GIS software from ESRI was used to analyze spatial data provided by the Homicide Task Force (HTF) during the investigation of the serial killings in Spokane. The Animal Movements software extension (available for download from the Internet), developed by Phillip N. Hooge at the US Geological Survey's Alaska Biological Science Center, was used to examine

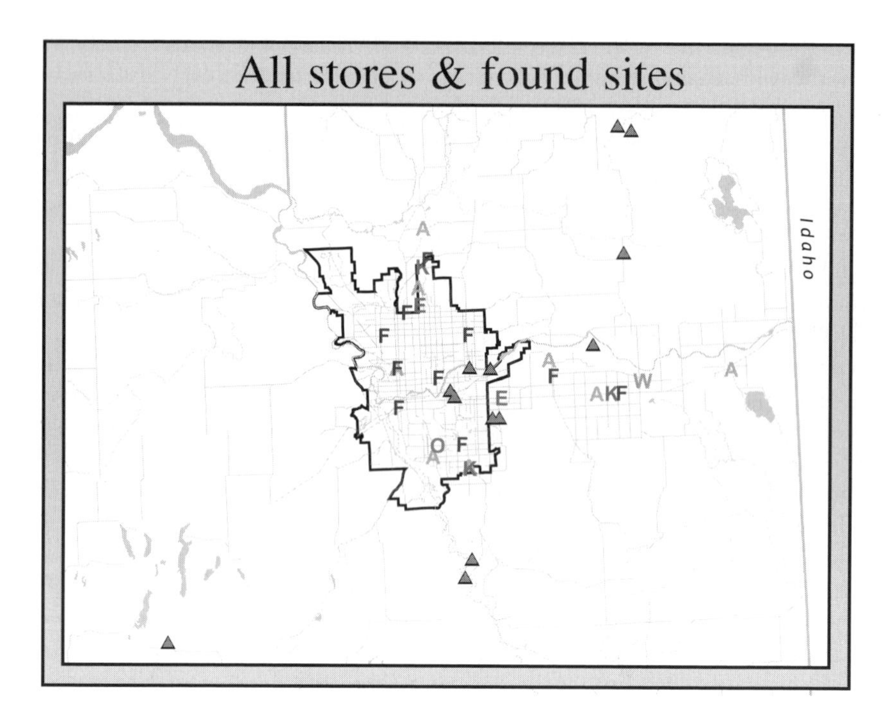

Figure 12.3 This map shows the locations (triangles) where the bodies of Robert Yates' victims were dumped. It also shows locations (depicted with letters) where supermarkets from which bags found over the head of Robert Yates' victims could have originated. A is for the units of Albertsons, E is for the units of Eagle, F is for the units of Safeway, W is for the Wal-Mart store in the area, K is for the Shopko units and O is for the Super One store. Note that there is only one Super One unit in Spokane. Note also that the cluster of stores is tighter than that of body disposal locations, a less frequented area being apparently preferred by the killer for disposal of his victims' bodies. Later analysis omitted the Wall-Mart store as it was felt it was both an outlier in terms of its location and not exactly in the same category as a neighborhood supermarket as were the other stores. In subsequent analysis Wal-Mart and Eagle were dropped because they were considered to be in a different class from routinely patronized grocery stores.

spatial relationships between the locations where victims' bodies were being dumped and the locations of potential sources of material evidence found on those victims. Specifically, there was one or more plastic bags found on the head of all but one of the victims; the bags came from a handful of Grocery stores in the Spokane area. Since there were multiple locations for many of the stores identified on the bags, the HTF asked to look at the locations of all the stores represented with respect to the locations of all the victims. The Animal Movements software provides several different algorithms for the

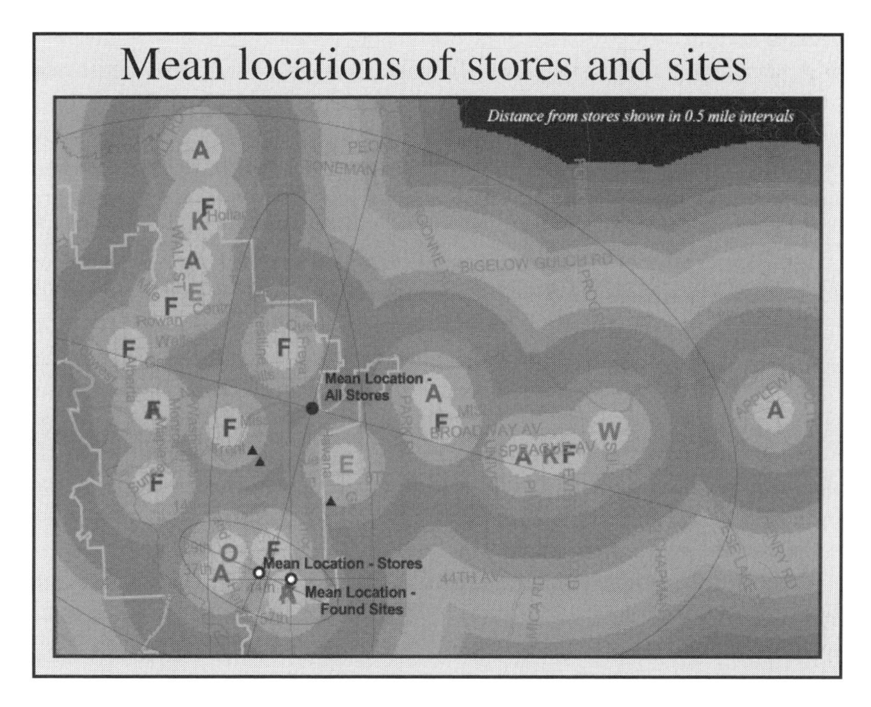

Figure 12.4 This map shows the results of the spatial analysis of the locations where bodies were found and stores where bags found with bodies may have originated. Note that the ellipses generated by the geo-statistical analysis have differing axis but that the mean location (centroids) of these ellipses both are centered on the same South Hill neighborhood and are less than a mile apart.

spatial analysis of point phenomena. The Jennrich-Turner (1969) Bivariate Normal Home Range method was used to calculate probability ellipses and arithmetic mean locations for the stores and the victims. A subset of six of the victims was then examined using the same methodology. The ellipse and mean location of the six victims pointed to a subset of the stores, which was then analyzed. The probability ellipse for the subset of stores overlapped that of the six victims showing an area in common of fewer than five square miles out of more than 1,800 square miles in Spokane County; moreover, the mean locations of the two subsets lay within one mile of each other. See Figure 12.4 for the probability ellipses generated by the first pass at spatial analysis of body and shopping center locations.

GIS methodology

1 The HTF supplied data on the locations of found victims, which were entered into the GIS using geographic coordinates that were re-projected

to produce a shapefile for mapping and analysis with the existing data sets.

2 A list of stores represented on the bags in evidence was provided by the HTF. The addresses of the stores were then obtained from the phonebook and the list of addresses was geo-coded to the county road file to produce a shapefile containing the locations of the stores within the study area.
3 The ARCView spatial analyst extension was then used to create a grid layer showing distance intervals from each store radiating out in half-a-mile increments. The locations and distances from stores with respect to found victims were then displayed on a map provided to the HTF.
4 The HTF then asked that spatial analysis be performed on the point data to calculate the mean locations of the stores and the victims. The Animal Movements extension was downloaded from a USGS Internet website at: www.absc.usgs.gov/glba/gistools to accommodate the analysis. After consulting the Animal Movements documentation, the method selected for performing the analysis was the Jennrich-Turner Home Range option. The resulting probability ellipses and mean locations were plotted on a map for the HTF.
5 A second analysis was then performed using a subset of the victims and a subset of the stores. The resulting ellipses and mean locations were again plotted on a map for the HTF showing an area of overlap in the probability ellipses that was located on Spokane's south hill.

Results of geo-spatial analysis

The results of the analysis indicated that the centroid for store locations was in the upper middle-class South Hill neighborhood. It was within a half mile of the centroid for the body disposal locations. The results of applying this refinement of the model are presented in Figure 12.5. One can see that the body dump locations and supermarket locations when analyzed using the model, each generates an ellipse and the axis of the ellipses are not aligned. However, the centroid of each ellipse is located in the same general area and both are in the South Hill neighborhood. Four months after this analysis was performed, the suspect who would be convicted for these crimes was discovered and apprehended. The discovery was made primarily on the basis of indications of suspicious activities by the driver of a white Corvette and as a result of a stop by a Spokane police officer of a white Corvette being driven by Robert Yates. Later, latent evidence recovered from the stopped Corvette (which by then had a new owner) revealed bloodstains that were linked to one of the victims. Although GIS and geographic profiling did not solve this case, in retrospect it could have narrowed down the focus of interest to a five square mile area out of 1,800 square miles within the greater Spokane metropolitan area. The suspect's home residence turns out to have been located within this five square mile "high probability" area; in fact he

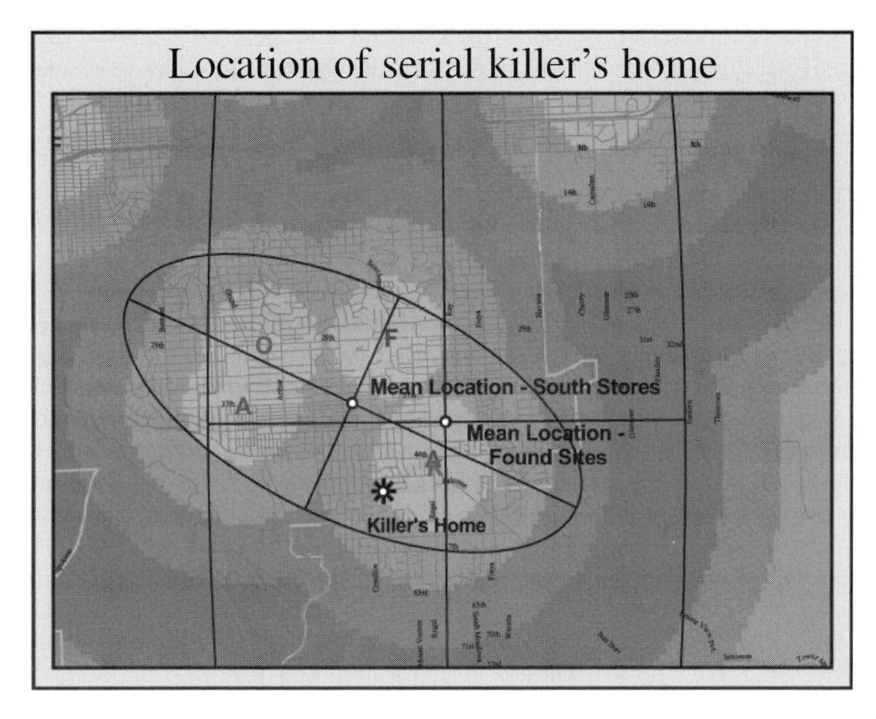

Figure 12.5 This map shows the South Hill focus area with the mean locations for the sites at which bodies were found and that sub-set of stores (excluding Wal-Mart) that were in that area of Spokane. The home location of the killer is also displayed. Thus the killer lived within the ellipse for both store locations and body disposal locations and less than one mile from the centroid for each of these clusters. This neighborhood occupies less than 1/10 of 1 percent of the greater Spokane area.

lived less than a mile from the mean locations (centroids) of both the body dump and the grocery store clusters. Since approximately 5,000 persons reside in this area out of a regional population of 420,000, that information coupled with other clues (how many of that subset of persons owned a white Corvette?) could have been very helpful to investigators. In this case, the approach to geographic profiling would seem to be a useful additional tool for investigators.

GPS data in the Yates case

The Robert Yates case is fascinating in itself as a study of an organized serial killer and as a demonstration of the utility of geographic profiling. The case nevertheless has a twist that may make it nearly unique in the annals of crime and its investigation. The twist is that Robert Yates was a frequent and apparently expert user of GPS technology. Yates had been trained as a forward observer in the US Army and there had been exposed to GPS technology and other aspects of orienteering. Yates had continued his affiliation with the Army as a member of the Reserves while living (and killing) in Spokane. Yates' use of GPS in his activities has a parallel with another more recent case involving a US serviceman. In this Virginia case, a former US serviceman with a high security clearance was arrested on suspicion of espionage. In his possession was a GPS unit, which allegedly marked locations of, suspected "drops" where his handlers from a foreign power would leave instructions or pick up classified documents.

When Yates was arrested, a Magellan 2000 XL GPS unit was recovered from his home. There are allegations that in a call from jail monitored by authorities he instructed his wife to destroy the unit. Although this allegation is unlikely to ever be admissible as evidence, it may have helped focus investigators interest on the unit. In any case, the unit contained 72 waypoints associated with several journeys in Washington State and adjacent areas of Oregon. Some of these journeys were made while flying a military helicopter. The reason authorities are interested in these waypoints is that they might possibly mark potential locations where bodies could be disposed of. Many of the locations are where isolated roads crossed rail lines or the flight path of the helicopter. With a unit like the Magellan, one need only select the "go-to" feature and the unit will provide continuously updated information on the distance and bearing and even estimated time of arrival at the previously marked location. For a serial killer trying to locate an isolated dump or gravesite such data would be invaluable. This would be particularly true if the perpetrator of a murder were traveling at night in an unfamiliar area. Of course other data on the routes that could lead to this point would be of assistance. But with the coordinates of the potential burial location, anyone with a modicum of navigation and map reading skills (such as are inculcated in army training exercises), could find their way to such a marked location. In order to investigate possible links between the waypoints stored in the GPS unit and potential crime scenes, the Spokane County GIS specialists enlisted the help of the manufacturer (Magellan) of the unit to extract GPS points stored on the unit. These were entered into an appropriate base-map (covering all of Washington State and Parts of Oregon). This data is providing numerous locations for field investigations. Helicopters are being used to assist the process of getting field teams to these locations. Based on data derived from this source as well other information of Yates' travels possible links to several other unsolved homicides have been established.

GPS data management methodology

The following passage describes the process undertaken to convert the GPS data points stored in that receiver and allegedly collected by convicted serial killer Robert Yates, into a GIS map layer. This work was performed by Spokane County GIS specialists for Detective Fred Ruetsch and Sergeant Cal Walker of the Spokane County Sheriff's Office.

Initially, Spokane County Detective Fred Ruetsch delivered a list of seventy-two waypoints in latitude and longitude coordinates (degrees, minutes, seconds) to GIS Specialist, John Bottelli of the Spokane County GIS program. The coordinates were then entered by John by keyboard into an excel spreadsheet. The columns were then flipped to longitude–latitude, adding a "negative" before each of the longitude coordinates and then the data was exported as a .csv file. Once this was completed the .csv was converted to a UNIX file and the coordinates projected into decimal-degrees using the Arc/Info version 7.2 command: PROJECT FILE. And this was ultimately re-exported as a UNIX file, and set as the INPUT file to generate a point coverage in ARC: with the GENERATE command. Then the coordinate system was defined for the newly created point coverage using the PROJECTDEFINE command in ARC and specifying the characteristics of the GPS source data. After the new coverage for points was built, the PROJECT command and a projection file were used to convert the point coverage into geographic NAD83 format file compatible with the Spokane GIS base-map. Next, the geographic NAD83 data points were projected into an Albers Conic Equal-Area projection. Finally, unique symbols were assigned to waypoints from each of three separate dates. The dates of interest to the detectives were: 5 June 1999, 13 June 1999, and 18 June 1999. This summary of the actually far more complex procedures involved in data conversion indicates the technical difficulty in converting raw GPS waypoints stored in a "recreational" grade GPS unit like the Magellan into a point coverage suitable for use in a GIS-based analysis.

In order to create a state-wide map of the GPS points, it was necessary to obtain GIS data from the web, since Spokane County's in-house base datalayers are limited to the County itself. Free downloadable statewide GIS data was obtained from the Washington State Department of Transportation and the Washington State Department of Ecology web sites. Among the data layers obtained were: counties, highways, cities, hydrography, railroads, graticule (latitude longitude lines), federal lands, rest stops, and park and ride locations. These layers were not chosen randomly. Rather they were either related to potential body dump locations (i.e. national forests), transportation access (highways), locations where victims might have been encountered (rest areas) and jurisdictional boundaries (county boundaries).

All of the various data layers were in differing projections or in geographic coordinates so a common system of geo-referencing was needed. In order to effectively show the GPS point data, which covers the whole of Washington State, an Albers conic equal-area projection was selected. Once all the various data sets were in a common map projection, a statewide e-sized small-scale map was produced displaying the GPS points in conjunction with all the data identified above. See Figure 12.6 for a state-wide

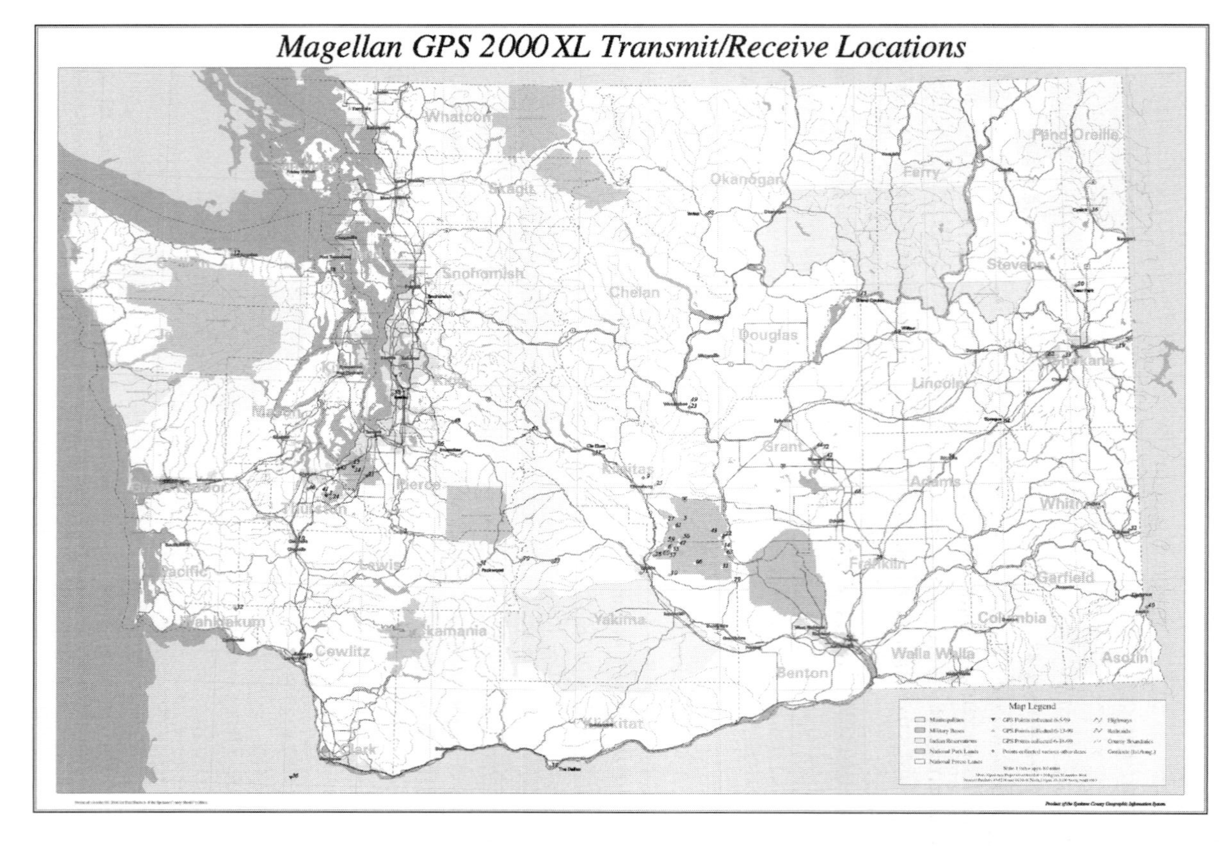

Figure 12.6 This map depicts GPS waypoints in Washington sate recovered from Robert Yates' own personal Magellan GPS unit. These provide investigators an indication on the travels of this prolific serial killer.

map of the tracks and locations of the seventy-two waypoints recovered from Robert Yates' personal GPS unit. This map was provided to the local TV and print media and provided to other law enforcement organizations throughout Washington State.

Status

Robert Yates pleaded guilty to ten murders in exchange for the prosecutor in Spokane not seeking the death penalty. Therefore no GIS products figured in his trial and in truth they would have only provided a spatially based indication of his involvement, while DNA evidence found in his white corvette and linked to an early victim was highly incriminating. Yates was sentenced to 408 years in prison. In addition, he is, as this chapter goes to press, being tried for two further homicides. Prosecutors in that case are seeking the death penalty. Data generated from GIS-based analysis of Yates' own travels as recorded in his personal GPS unit may figure as evidence in these cases in which he is contesting his guilt. Whether it is persuasive or not remains to be seen. Nevertheless, in the Yates case as in the Jackson Case, GPS and GIS played a role. These cases show that geo-spatial technologies, once solely limited to military, natural resources and infrastructure management applications, are permeating society and affecting both the enforcers of – and offenders against – the law.

13 The Enforcer GIS

Helping Pinellas County, Florida manage and share geo-spatial data

Tim Burns, Mark R. Leipnik, Kristin Preston, and Tom Evans

This chapter underscores the importance of sharing data across largely artificial jurisdictional boundaries in achieving efficient law enforcement. In particular sharing geospatial data is important in comprehending and responding to the challenge posed by crime in increasingly urban areas with multiple jurisdictions. GIS can serve as a mechanism for not only sharing data among various agencies, but also uniting efforts in combating crime. This chapter also points the way for the many Sheriffs' Departments (and other regional law enforcement agencies) to the benefits of GIS as well as the issues related to adopting GIS.

(Editor)

Introduction

Pinellas County, Florida is a rapidly growing county on Florida's Gulf Coast slightly east of Tampa. It is the home of the world famous resort cities and the retirement "Meccas" of St. Petersburg, Clearwater, and Tarpon Springs. As a popular residential and tourist location, Pinellas is one of the most densely populated counties in Florida, ranking fifth in population, with approximately 921,000 full time residents located within the county's 280 square miles. This population is spread across twenty-four municipalities and large unincorporated areas varying greatly in population, size and economic status. Coupled with a dense population and municipal divisions, the county contains twelve municipal law enforcement agencies and a large Sheriff's Office each with distinct, intertwined jurisdictional policing boundaries. This diverse policing authority, sharing many common borders, implies a need to share data to encourage proactive policing and efficient resource allocation within jurisdictions. (See Figure 13.1 for a county-wide jurisdictional map).

To better deploy resources, analyzed crime patterns and generally manage data and resources throughout the county, the Pinellas County Department of Justice Coordination has adopted the use of geographic information systems (GIS) technology via the creation of a system cristened "the Enforcer" GIS. The Pinellas County Department of Justice Coordination is a policy, research, and project management agency residing under the County

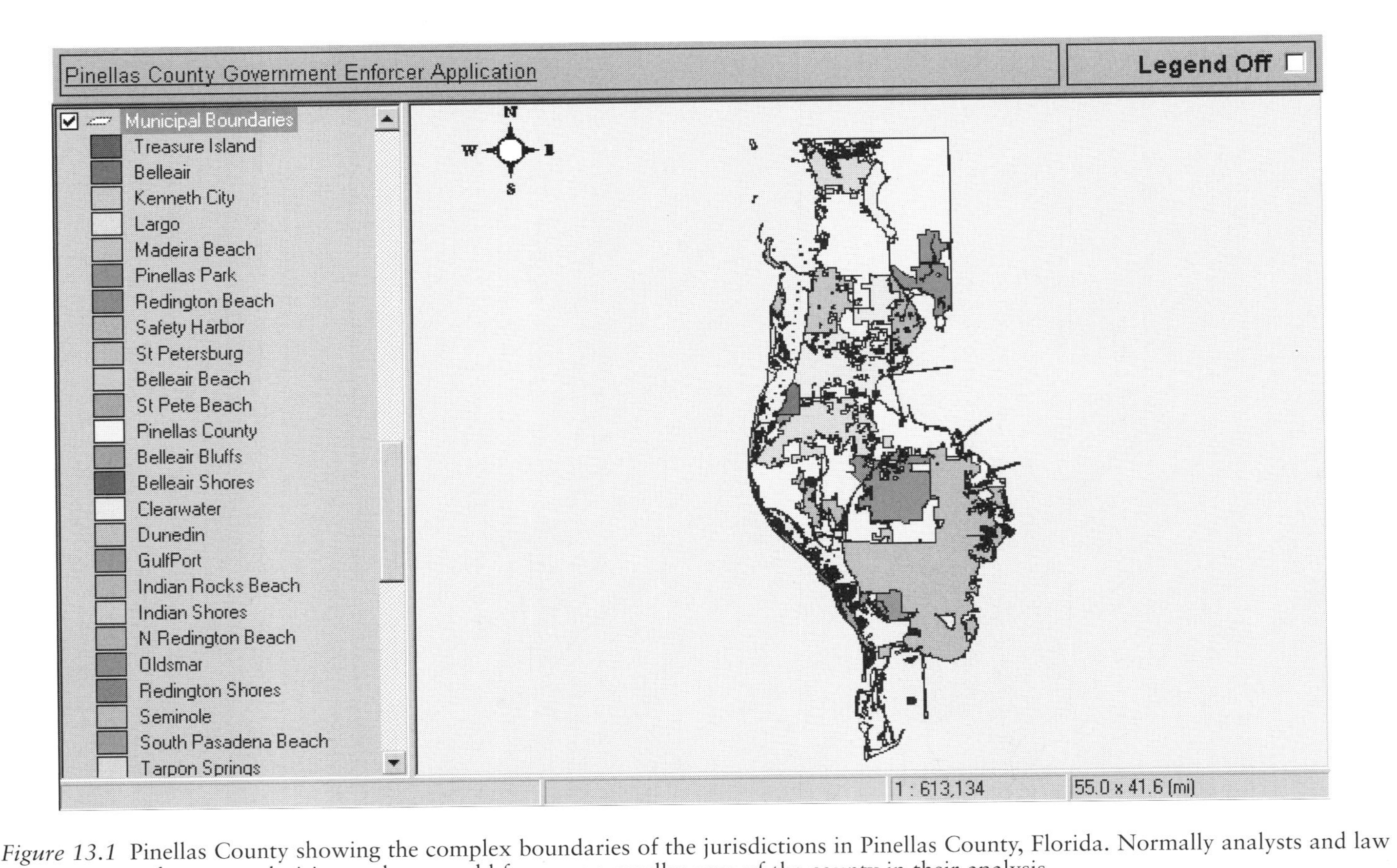

Figure 13.1 Pinellas County showing the complex boundaries of the jurisdictions in Pinellas County, Florida. Normally analysts and law enforcement decision makers would focus on a smaller area of the county in their analysis.

Administration, serving the needs of the Board of County Commissioners. This department provides a key role in supporting both internal and external county justice initiatives. Some of the services the department provides are assistance with information gathering, policy analysis, project coordination, grant writing and facilitation, and contract management. Within the Enforcer GIS Project, the department acts as the central, impartial facilitator to accomplish project goals.

The GIS is unusual in using a highly accurate street and parcel base-map (see Figure 13.2) with easy access to numerous other county-wide layers. It is also unusual in that it utilizes GIS software from AutoDesk, Inc. along with the Oracle relational database. As such it is one of the most advanced county-wide law enforcement GIS systems in the United States. Relatively few of the approximately 3,141 counties (some cities function as quasi-counties, in the US making determination of an exact number problematic) have as of yet adopted GIS technology, perhaps because their patrol responsibilities are often in rural or semi-rural areas. Thus, Sheriff's Departments have lagged municipal police departments in GIS use. The Pinellas County Department of Justice Coordination, the County Sheriff's Department and the other participants in the Enforcer GIS, should serve as a model for many other counties, and regional law enforcement agencies generally. The following material provides an overview of the salient features of the Enforcer System and examples of representative crime-related maps generated using the system.

Enforcer overview

The Enforcer GIS is an enterprise-wide initiative originating from the Pinellas County Department of Justice Coordination. It is an attempt to establish data sharing and analysis tools between agencies. The project seeks to pursue the use of complementary technologies in order to allow for proactive policing and problem solving across jurisdictional boundaries. These technologies include:

- a Geographic Information System (GIS)
- an integrated relational database with automated data management
- a web-based crime mapping interface to provide decentralized access by any officer and
- infrastructure technologies for security and efficiency to eliminate system access barriers in the office, in the field, or even at home.

The Enforcer GIS is intended to improve the ability of the county's various law enforcement agencies to:

- monitor and analyze criminal activity in, and in areas surrounding, their own jurisdictions
- improve each agency's ability to recognize and respond to evolving crime patterns

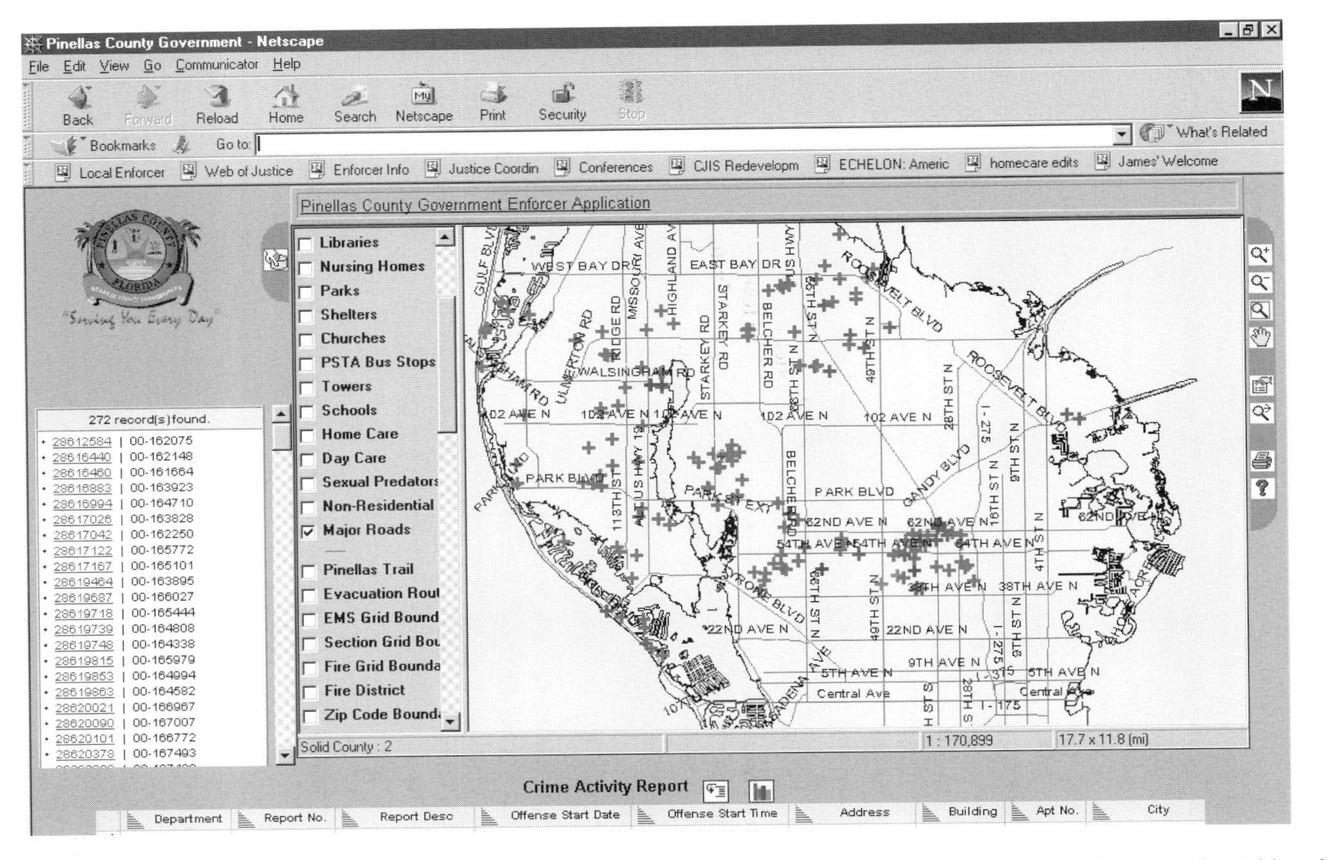

Figure 13.2 Typical product generated interactively over the Internet by a user of the Enforcer GIS. This map shows residential burglaries for September 2000 for one of the participating municipalities. The major streets layer is turned on but the parcel map is not displayed.

- enhance regional operational planning and
- improve regional decision-making.

The system also enhances the information each agency has available upon which to base decisions concerning the allocation and deployment of resources for the prevention and suppression of criminal activity. The system is not meant to replace the activities currently conducted within agencies, but to be a new tool to supplement each agency's resources.

Project development has been based on Pinellas County's Geographic Information System which currently contains hundreds of well maintained layers and updated, geographically-referenced attribute information, stored within an Oracle relational database. It has been developed through the use of Systemhouse trademarked Vision Software. The county GIS has been a cooperative effort operated and maintained by the Board of County Commission Information Systems Department (BCCIS). It has grown over the years to include data from the property appraiser, planning, supervisor of elections, emergency management, development review services, general services (real estate division), social services, community development, public works (highway, solid waste, traffic), environmental services, utilities, and even animal control services with more partnership efforts planned for future development. Using this system as the basis to the Enforcer GIS allows each agency to work on the same standard data set when analyzing crime.

Governance and participation

The project maintains a participatory governance structure in order to allow each active agency an equal voice in project direction and development. This governance structure includes an active Steering Committee with one command staff representative of each agency that meets on a quarterly or as-needed basis to decide upon project priorities. An operational committee exists in order to ensure that day-to-day activities are completed and issues addressed. Additionally, the Steering Committee has the authority to call for the establishing of special issue subcommittees that will be directed to explore or complete specialized tasks. Along with beginning this initiative, the Department of Justice Coordination has taken the active role of facilitating the project development and participation. Justice Coordination owns both the hardware and the interface. However, by agreement, all data that is entered into the Enforcer System remains the property of the originating agency. If, for whatever reason, an agency wishes to withdraw from the project, an agency can terminate participation with written notice and have their data eliminated from the system.

Current status

The Enforcer Project has evolved into a successful program between the Pinellas County Department of Justice Coordination, the Pinellas County

Sheriff's Office, the Florida Department of Corrections, the Pinellas County Licensing Board, and municipal law enforcement agencies. The latter consists of eleven police departments (Clearwater Police Department, St. Petersburg Police Department, Gulfport Police Department, Pinellas Park Police Department, Kenneth City Police Department, Belleair Police Department, Largo Police Department, Treasure Island Police Department, Tarpon Springs Police Department, Belleair Beach Police Department, and St. Pete Beach Police Department) with system support provided by the BCCIS. Future expansion is geared towards partnering with additional agencies, addressing additional data needs, and providing enhancements to the system functionality.

The Enforcer System model uses a central database with data being updated from each agency's system to the Enforcer. The project goals are geared towards automation of this process to allow for hands-off updates. Currently, data from the Pinellas County Sheriff's Office, which includes data from all unincorporated areas and twelve contract municipalities, have been entered into the system in an automated form on a daily basis for access by other agencies. Along with the Sheriff's data, the project has been working with the integration of data from Pinellas Park, Gulfport, Belleair, St. Petersburg, Clearwater, and Largo, with plans to begin work on data from Kenneth City, Treasure Island, Tarpon Springs, and St. Pete Beach, utilizing a recently awarded grant. Additionally, sexual predator information is being entered into the system by the Department of Corrections for law enforcement use, and the Pinellas County Licensing Board has been providing information on all home care centers and day care centers to the system through a separate web interface (see Figure 13.3).

System access

The Enforcer System administrators, in partnership with AutoDesk, Inc., has designed a user-friendly, web-based interface for access to system data and to perform simple crime mapping. This interface will provide a common platform for all agencies to access the system. The Enforcer interface is based on AutoDesk's dynamic, web-based Mapguide product with Phase I delivery accomplished on October 1, 1999. This interface has provided an innovative vehicle for any officer to gain access to crime mapping and information through a common web browser (i.e. Netscape or Internet Explorer). See Figure 13.2 for the interface and a typical map generated using the system. Recent development by AutoDesk, Inc. and Focus Advanced Technologies has provided a Phase I Version B upgrade to the original interface with additional graphics and dynamic functionality that will be implemented over the next few months. Phase II interface development discussions have begun with AutoDesk for sophisticated crime analysis additions including hotspot analysis, predictive modeling, forecasting, and correlation analysis. Also, new system uses and access have been

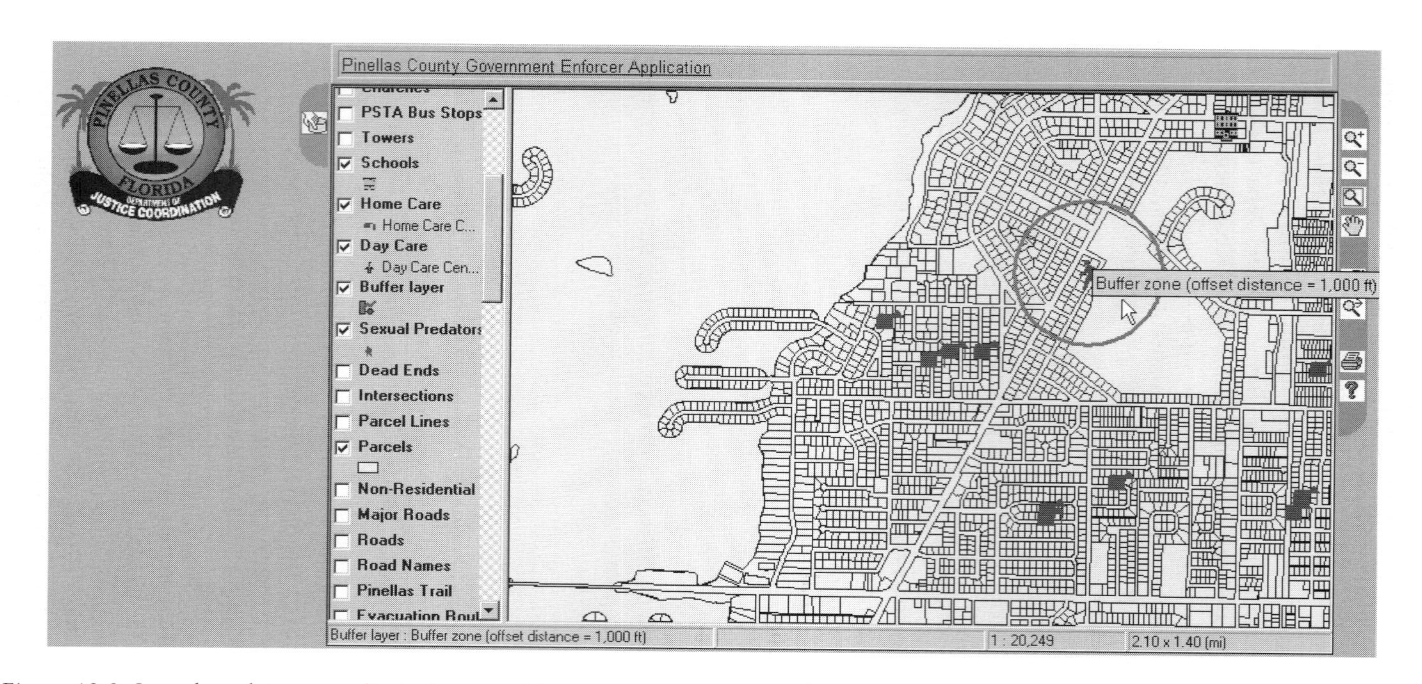

Figure 13.3 Sexual predator map displaying parcel base, home care centers, day care centers, schools, and residence of a sexual predator with 1000 ft buffer zone generated around it. The existence of a link to a cadastral database facilitates using home address information to accurately geo-code offender residence information, this in turn is important in determining compliance with exclusion zones around facilities that house children vulnerable to such predators.

discussed and will be pursued as current system priorities are achieved. Some of these ideas include floorplans for emergency management, central pawn information management, gang tracking, auto theft management, and much more. Through focusing on a modular development of applications, such as sexual predator management, crime mapping, and analysis, the Enforcer GIS has been able to continue to provide incremental enhancements within common, user-friendly framework.

The system resides within the Pinellas County Intranet to allow for system security and users are password-authenticated for access. For those agencies without a dedicated connection to the Pinellas County Intranet, system, delivery will be achieved over the Internet through the use of Virtual Private Network (VPN) and Citrix Independent Client Architecture (ICA) technologies. The VPN will allow a user to be authenticated into the Pinellas County Intranet through any Internet Service Provider connection creating a secure tunnel over the Internet for information to be transferred. The ICA technology will allow a user's system to act as a "thin client" when using the Enforcer allowing for great savings in bandwidth and much increased efficiency of the system. The application being used over ICA will have the exact same look and feel. However, it gains its efficiency through leveraging the server to actually provide the processing of the program while only sending mouse clicks, keystrokes, and images to the user's PC.

Summary of system benefits

The key benefits of this system include:

- Improvement of cross-jurisdictional data sharing.
- Improvement of the ability to recognize and respond to evolving crime patterns along jurisdictional boundaries.
- Enhancement of regional operational planning.
- Achievement of valuable time and money savings through improved information and analysis.
- Elimination of the cost and expertise barriers to data sharing and GIS.
- Enhancement of resource allocation (better deployment of officers).
- Facilitation of proactive community-oriented policing.
- Achievement of time savings in crime analysis, thus allowing analysts the time to address more complex crime issues.
- Overall enhancement of services and protection of citizens.

Each participating agency can provide access to any number of their officers and command staff, in order to easily see crime patterns and proactively acquire information. With the GIS in place, a multitude of additional information (layers) is made available for law enforcement use; including parcel information with land use, owner information, and sale histories, emergency shelter information, evacuation zones and routes, false color digital ortho-rectified aerial imagery and many other layers and associated attribute data.

In the Pinellas County Sheriff's Office, the Enforcer GIS is used at all levels of command and support. The management staff is able to use the Enforcer to identify hot spots and query the system for decision support and resource allocation based on incident reports. The query can be based on yearly, quarterly, monthly, day of week, and time of day incident reporting features. With this easy access, the system becomes a kind of manager's desktop *Comstat* application. Detectives can view where particular crimes are occurring and community-policing staff can look at specific neighborhoods and subdivisions. Support personnel can select any number of parcels to create a listing for mass mailings along with maps to be sent out to the community.

Another example of an innovative use of the Enforcer System has been sex offender management within Pinellas County. The Florida Department of Corrections (DOC) has been using the system to enforce state statutes and regulate where sex offenders can reside within the county. Additionally, law enforcement is required to provide notifications to all of the home care centers, day care centers, and schools within one mile of the residence of a registered sex offender. The Enforcer provides a central location to manage this data, allowing users to generate buffers around residences, and generate lists of day care, home care, and schools with addresses. This cooperative effort takes school locations from the county's land use information, home care and day care information updated through a web-enabled interface from the Pinellas Licensing Board, offender information from DOC, and various other layers from our system to provide a tremendous time saving tool (see Figure 13.3). In the past, by the time probation officers checked out an address where an offender was living, they then had to go to court for a judge to make the decision on whether they had to move or not. Today, the process is proactive letting the offender if a potential residence is acceptable know before the move occurs.

Funding and recognition

Funding for the hardware, data integration, and interface licensing has been acquired through grants generated by the Department of Justice Coordination. Justice Coordination will continue to pursue grants in order to further system development.

The Enforcer System has been modeled as a compilation of innovative concepts and technologies. The result of this model has brought a great deal of recognition to the project in the form of articles, news broadcasts, conference presentations, and visits by interested parties and this chapter.

Agency participation

The Enforcer Model is designed to minimize the efforts required by participating agencies. The project plan is to hire a third party integrator to provide an automated integration of agency data into the system. The key responsibility of each agency will be to work with the integrator to provide a data extract in a specified format. The integrator will provide scripts for data transport, conversion, and uploading.

It is important for the Enforcer Project to encourage the participation of all law enforcement agencies within the county. The result of data not being available from a jurisdiction can greatly impact the way the system can be used. Missing an agency as large as the St. Petersburg Police Department would create a large void in the information for Southern Pinellas County. Figures 13.4–13.6 demonstrate the importance of establishing a cross-jurisdictional system for data sharing and analysis. Without the partnership effort shown by each participating agency, a cross-jurisdictional project such as the Enforcer GIS Project would not be possible.

In order to further encourage agency participation within the Enforcer Project, Justice Coordination has been providing yearly training courses to provide a common, cross-jurisdictional understanding of effective data utilization. As a courtesy, in conjunction with the Enforcer Project, Justice Coordination funds a closed training course once a year and allocates seats to the training to all law enforcement agencies within the county. The courses in Crime Analysis Applications, Criminal Intelligence Analysis, Statement Analysis, and Criminal Investigative Techniques have proven very effective, with one agency solving a high-level robbery using the acquired techniques.

Additional information about the Enforcer GIS Project can be obtained by contacting James Dates, the Department Director or Tim Burns, the Justice Information Analyst at the Pinellas County Department of Justice Coordination, 311 S Osceola Avenue Clearwater FL 33756 or see the Department website at: http://www.co.pinellas.fl.us/bcc/juscoord/enforcer.htm

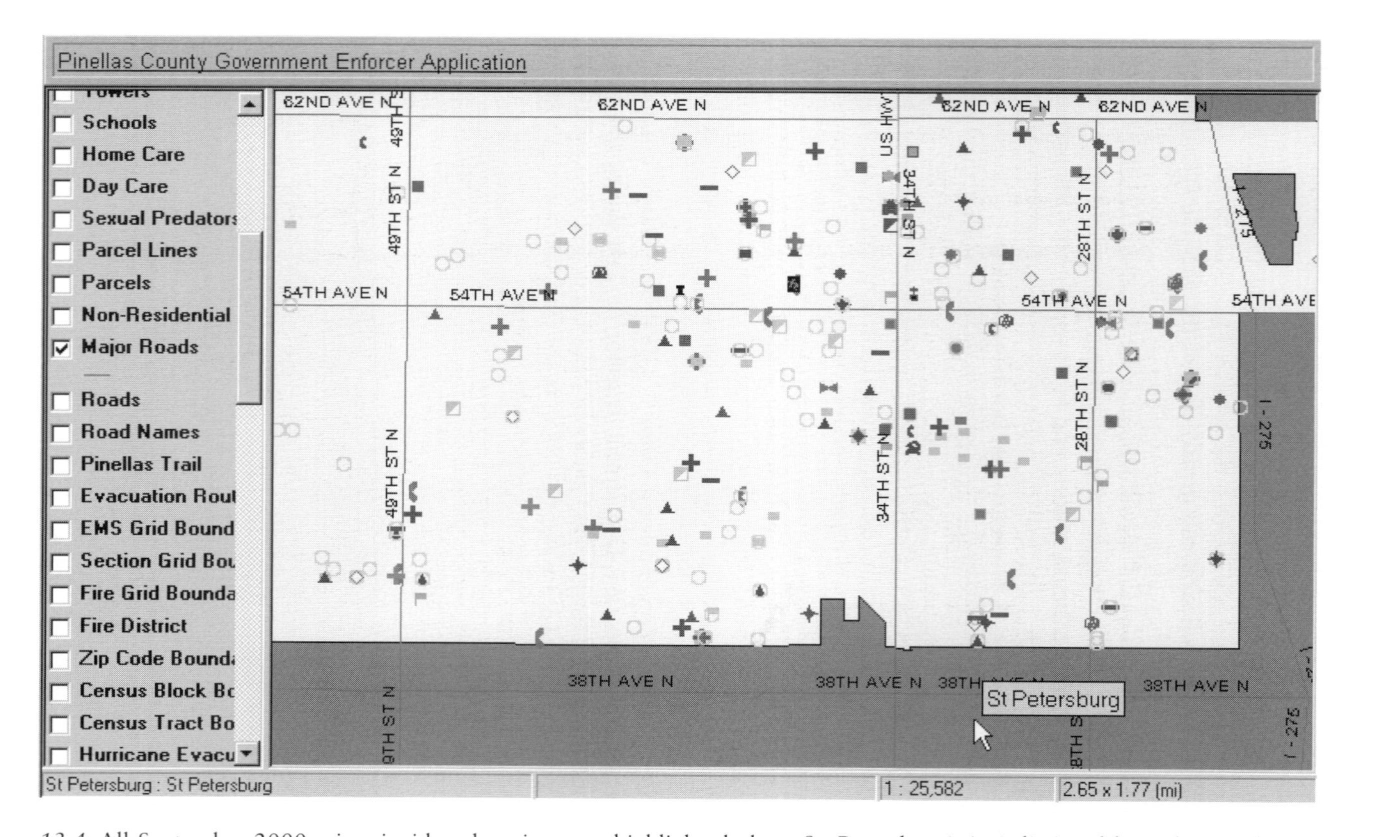

Figure 13.4 All September 2000 crime incident locations are highlighted along St. Petersburg's jurisdictional boundary with unincorporated areas of the county. Note that no incidents within the City of St. Petersburg are displayed.

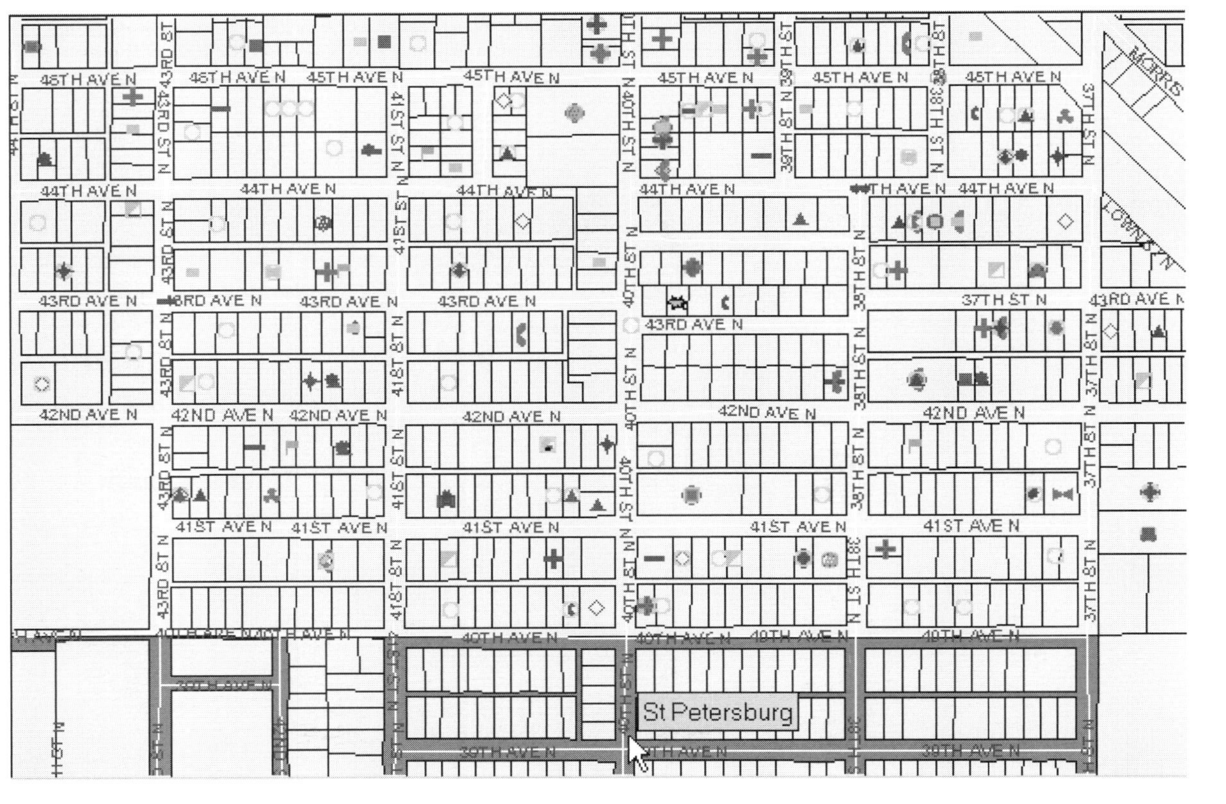

Figure 13.5 All September 2000 crime incident locations displayed for the area along the boundary of the unincorporated area of the county with St. Petersburg's jurisdiction. This map includes the parcel boundaries.

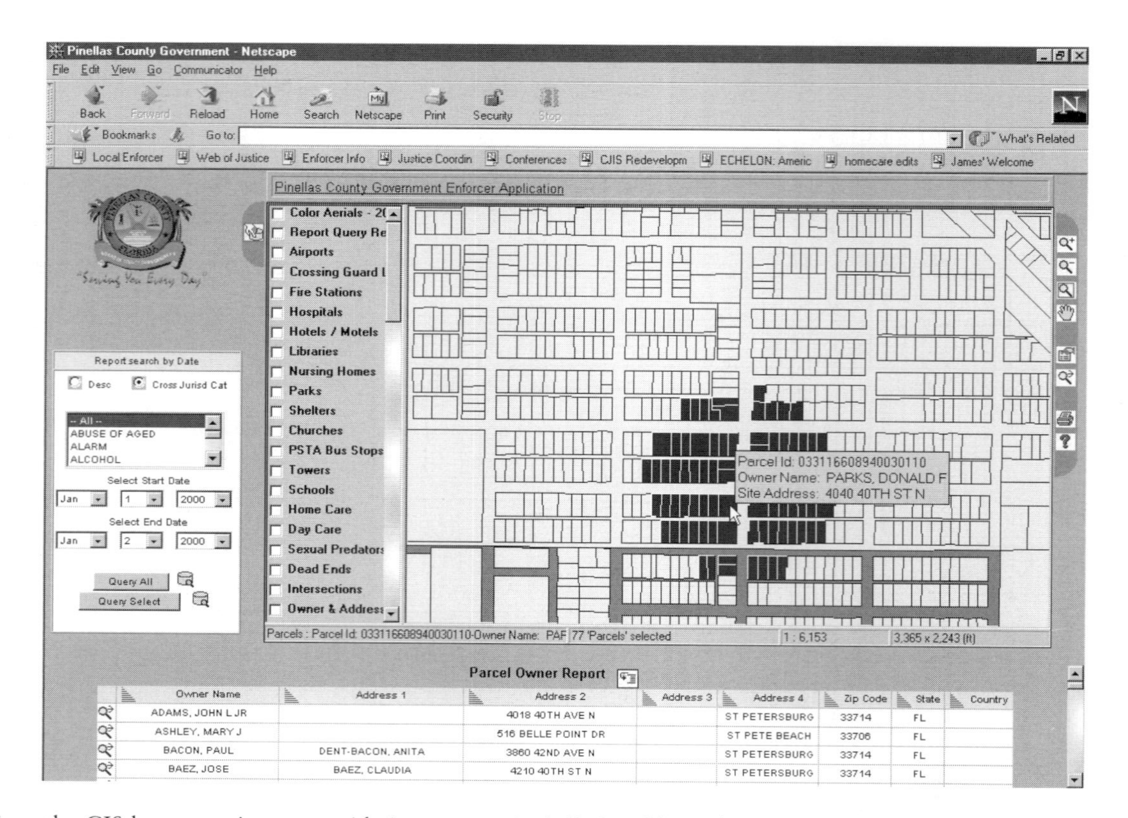

Owner Name	Address 1	Address 2	Address 3	Address 4	Zip Code	State	Country
ADAMS, JOHN L JR		4018 40TH AVE N		ST PETERSBURG	33714	FL	
ASHLEY, MARY J		516 BELLE POINT DR		ST PETE BEACH	33706	FL	
BACON, PAUL	DENT-BACON, ANITA	3860 42ND AVE N		ST PETERSBURG	33714	FL	
BAEZ, JOSE	BAEZ, CLAUDIA	4210 40TH ST N		ST PETERSBURG	33714	FL	

Figure 13.6 Since the GIS base-map is countywide it can cross jurisdictional boundaries in the course of generating maps or performing analysis. Here addresses of residents within a buffer zone can be extracted for interview or notification purposes regardless of whether they reside inside or outside the city of St. Petersberg.

14 The Delaware real-time crime reporting system

A statewide and enterprise-wide GIS application for law enforcement

Mark R. Leipnik and Donald P. Albert

> ... I want us to establish a statewide crime tracking system where crimes are immediately mapped and tracked so that all police departments statewide can have crime data on a real-time basis to better deploy their assets. With the State leading the way, I challenge Delaware's Police Chiefs to develop a plan by the end of March to implement such a system ...
>
> (Governor Thomas Carper, State of the State Address 1/22/98)

Statewide use of GIS technology is not unusual in application areas such as natural resources management or by State Transportation Departments, but it is highly unusual for law enforcement agencies at the state level to employ GIS comprehensively. This largely reflects the decentralization of law enforcement activities in the United States. In particular, very few states give local law enforcement powers to state agencies. Thus, the State Police are likely to be responsible for enforcement of laws and traffic regulations on state, US and Interstate highways and for special situations such as investigation of public corruption. A few states are using GIS to aid the highway patrol function (Illinois and New Jersey, for example) or in monitoring of parolees (Wisconsin and Texas, for example) but very few indeed are applying GIS to a wider range of law enforcement issues. An exception is the State of Delaware, where the State Police have local law enforcement responsibilities in most of the counties for those areas outside incorporated cities. This fact, plus the relatively small size of Delaware, has facilitated the creation of a wide area network-based statewide crime mapping and analysis capability.

(Editor)

The State of Delaware has developed a uniquely comprehensive enterprise-wide and statewide crime mapping, analysis, and reporting capability. Spearheaded by the Department of Public Safety (DPS), in Dover, Delaware, this project has over the last few years created a statewide GIS for crime mapping and analysis purposes. This GIS covers all forty-three jurisdictions in the state with local, county, or state law enforcement responsibilities. Delaware, the second smallest state in the nation, is 1,982 square miles in area. The state consists of three counties with a total population of about

760,000, much fewer than many cities and counties in the United States. The law enforcement branch for the state is the Department of Public Safety – Division of State Police (DSP) and is run by a superintendent with nine regions patrolled by troops. Crime has been on the decline over the past few years, with a 7 percent reduction from 1997 to 1998 and another 7 percent reduction from 1998 to 1999.

Of the forty-three police agencies in Delaware, thirty-nine are police departments in incorporated cities including Wilmington (the most populous city), and one is a county (New Castle) that has a police department of its own. The remainder of the state is within the jurisdiction of the State Police. In Delaware, the State Police have the unique responsibility of providing local law enforcement in all unincorporated areas outside the municipalities and one county. In addition to providing traffic enforcement on state and county roads, the Delaware State Police also provide community policing, preventative patrols, and responds to and investigates rural and suburban robberies, burglaries, family disputes, and all other manner of felonies and misdemeanors. Because of this unique responsibility, the State Police need some unusual tools and approaches. For example, the State Police have a fatal traffic accident reconstruction team that works statewide.

System development

With respect to the need for access to current geo-spatial information about the occurrence of crime, the DPS undertook the development of a GIS-based crime mapping and analysis capability that would be enterprise-wide and available over computer networks to state and local law enforcement agencies and offices (troops). The result was the creation of a Real-Time Crime Reporting (RTCR) system that is available on the state's Intranet for law enforcement officials only (Figure 14.1). This monumental effort involved the cooperation and collaboration of law enforcement practitioners and technology experts from across half a hundred agency and jurisdictional boundaries. The RTCR system was the answer to a previously stilted, slow business process of managing crime data. Throughout the state, officers wrote paper reports, the local agency either entered the data or sent it to the State Bureau of Identification for entry, and then it was entered into the state mainframe system. Reports, analyses, and maps were created by hand and distributed to the appropriate personnel. The average time to complete this process was three months.

In order to have a successful crime mapping and analysis system, the Department of Public Safety and the project team started with a number of goals for the system that included: improvement of the quality of life in Delaware by improving the quality of policing, fighting crime through strategic resource deployment, and to facilitate the two previous objectives to make very current data available on crimes occurring at all locations in the state. The DPS felt that this data should be available to all law enforcement

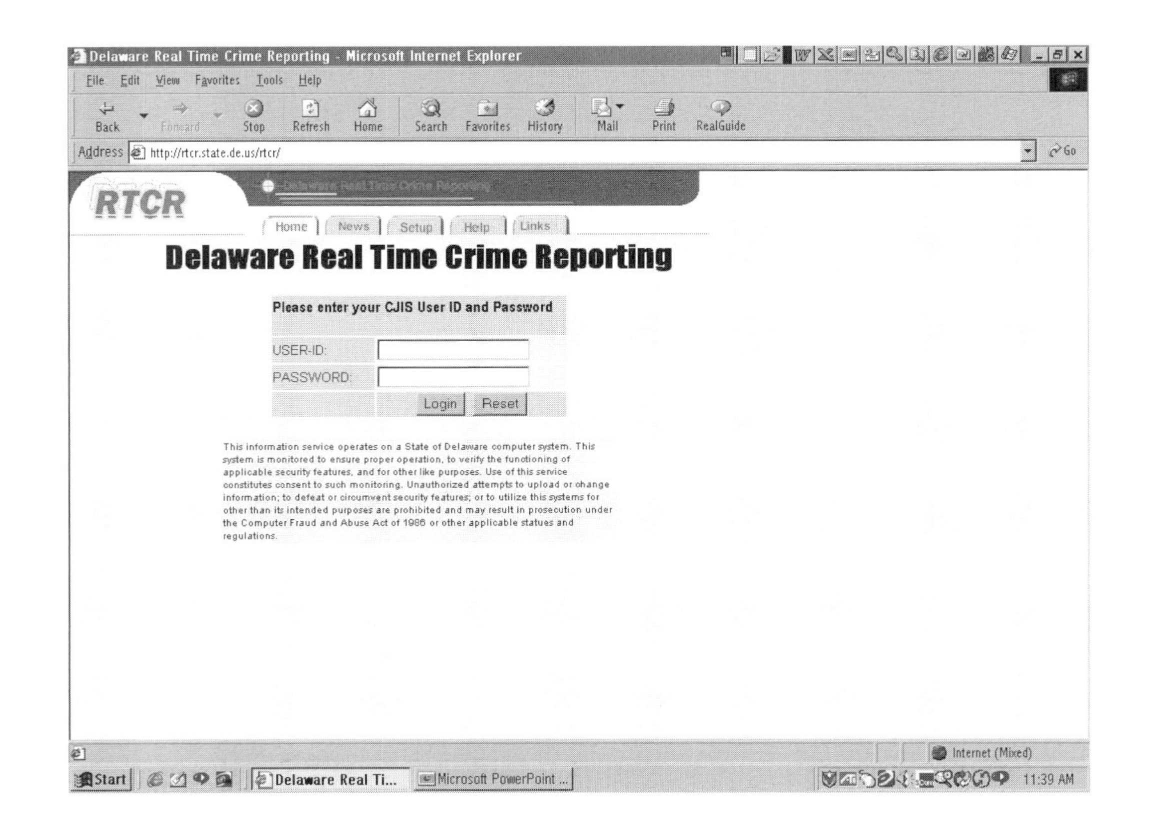

Figure 14.1 Log-in screen for the Delaware Real-Time Crime Reporting System, an intranet application that allows real-time crime mapping. Passwords restrict access to authorized personnel from the forty-three Delaware law enforcement agencies participating in the system.

agencies in the state. Therefore the DPS also made the important and ultimately vindicated decision to include the remaining areas (towns and New Castle County) into a single comprehensive effort spanning the entire state. Although Delaware's relatively small size undoubtedly made this effort less onerous than for many larger states, the benefits that all the law enforcement agencies in Delaware have derived from having a common base-map and common access over the Intranet to statewide real-time geo-spatial crime information in a uniform format are undeniable.

A key to the success of the RTCR system was the acceptance of the technology by its many users and stakeholders. Representatives from the various users and stakeholders were brought together to form two teams, one technical and another advisory, to guide the DPS through needs assessment, system design, and implementation. The technical team had representatives from Wilmington Police, New Castle County Police, Delaware State Police, Department of Transportation, Office of Information Services, and the Department of Public Safety. The Advisory team comprised of members from the Delaware Police Chiefs' Council, State Bureau of Identification, Delaware Justice Information System, State Planning, Statistical Analysis Center, and SEARCH – The National Consortium for Justice Information & Statistics.

In the needs assessment and system design process a conscious attempt was made to identify and solicit user needs from all stakeholders. Commanders and officers alike have felt they were part of the process from the start. Many details were incorporated into the system during the development phase, and several overall guidelines were agreed upon during a needs assessment and requirement study. These included:

1 that the system would be map-based and have an interface easy enough for officers to use with minimal training;
2 precise crime locations would be annotated on the map with details accessible in an attached database;
3 the system must be expandable to include accident and traffic summons data as well as data from other agencies;
4 crime data must be ready for analysis less than 12 hours after reporting;
5 the system must be accessible via Intranet or Internet; and
6 the RTCR must integrate with other automated policing systems.

Delaware awarded the implementation contract to a team comprised of Paradigm4 of Fairfield, NJ, Enterprise Information Solutions (EIS) of Columbia, MD, and Futurtech of Wilmington, DE. EIS specializes in GIS consulting and carried out the actual project implementation. Futurtech conducted the needs assessment, while Paradigm4 served as the project manager. After examining performance requirements, EIS and Public Safety selected the GIS development software that would function at the core of the system. Although off-the-shelf crime reporting GIS products are available,

Delaware wanted a customizable package and chose GeoMedia software from Intergraph Corp. of Huntsville, AL.

System design

The implementation team created a uniform statewide base-map derived from Delaware State Department of Transportation Intergraph files which – unlike those for most states – included local and arterial roads as well as state highways and Interstates. To this base-map were added the jurisdictional boundaries for the Police departments. The base-map had been created in an Intergraph MGE format (a format used by 48 out of 50 State Departments of Transportation but rare in law enforcement). A decision was made to utilize the Intergraph Geomedia GIS from Intergraph Corporation in Huntsville, Alabama. This GIS allows easy integration of non-topological data as well as accepting data from other GIS software such as ArcView into its open architecture structure. According to Eric Swanson, formerly the EIS Project Manager, "Another important advantage was that GeoMedia supports Active CGM technology. Delaware mandated a screen refresh rate of only 12 seconds, and Active CGM is the only way to accomplish that." In a client-server GIS environment, large vector map files require long download times. Every time the user wants to zoom in or pan out on the map, the map files must be reloaded from the server. Active CGM, a software tool developed by Interact Commerce Corp. of Scottsdale, Arizona, eliminates the reload step and lets the user move around the map without waiting for the screen to refresh.

The open architecture of the GIS also left Delaware free to choose virtually any commercial database system to store and retrieve the crime report data. The participants selected Oracle Corporation's Oracle 8 product with the Spatial Cartridge option. This product was chosen for its quick data retrieval capability and full compatibility with GIS environments. In early GIS systems, spatial information such as base-maps, transportation layers and aerial photographs were stored in separate databases from the attribute data that described them. This made data retrieval a time-consuming process. The new Oracle design maintains spatial and attribute data in one database, offering much greater efficiency.

Creating and populating the RTCR

The first step in building the GIS was purchasing a digital base-map. Because addresses would be used to locate crime scenes, the Department of Public Safety selected an off-the-shelf transportation map from GDT Inc. of Hanover, New Hampshire. This contained the layout of highways, roads and streets for the entire state. Establishing address ranges for the DOT base-map has proven a challenge; enhanced TIGER (topologically integrated geographic encoding and referencing) data obtained from GDT has been of considerable assistance in this effort. Then the Department obtained street name and address range data from emergency 911 centers in each of the state's three counties. Embedding the address ranges in the base-map was critical because it enables the GIS to display crime incident locations by street address precisely on the map display. Elayne Starkey, who headed up the implementation effort for DPS, made accurate geocoding of incident locations a top priority of the project.

Many Delaware police cruisers are equipped with GPS receivers for automatic vehicle location as part of E911. But these are mounted in the vehicles and cannot be carried directly to the crime site, which may be in a building or somewhere inaccessible to the cruiser. So rather than rely on GPS to locate the scene, the department decided it was more practical to use street addresses. Data entered into the electronic complaint form by the officer is automatically transmitted to update the Delaware Justice Information System (DELJIS), an existing comprehensive database of statewide crime reports. EIS linked GeoMedia directly to DELJIS and set up a computerized routine so that once every 10 minutes the details of new crimes would be downloaded to the RTCR Oracle database. Victim and suspect information never leaves DELJIS, but the location, type, and time of the crime are transferred to RTCR along with the reporting officer's name. RTCR tracks all Part 1 and many Part 2 crimes (Figure 14.2 and 14.3). In total, twelve different crimes are reportable in the system, but more will be added as it evolves. The current process has officers entering reports on their mobile data computers running a customized application called Enhanced Police Complaint (EPC) software. The reports are transmitted via cellular digital packet data (CDPD) wireless technology to a central database (DELJIS) where triggers send the selected fields of the crime data to the RTCR system. The data is then geo-coded and stored in Oracle Spatial and available within minutes for mapping and analysis over the network.

Hardware and software employed

A number of pieces of hardware are necessary to run this statewide system. The application is web-based but only available on Delaware's secure Intranet. Hardware includes:

- Two clustered Dell Poweredge 6300 servers
- Clarion C1500RA RAID 5 disk array
- 3Com SuperStack II Hub500
- Cisco 4000 Router
- Data Service Unit/Channel Service Unit (DSU/CSU).

In addition, officers are using mobile computers from Panasonic, Amrel, and PCMobile.

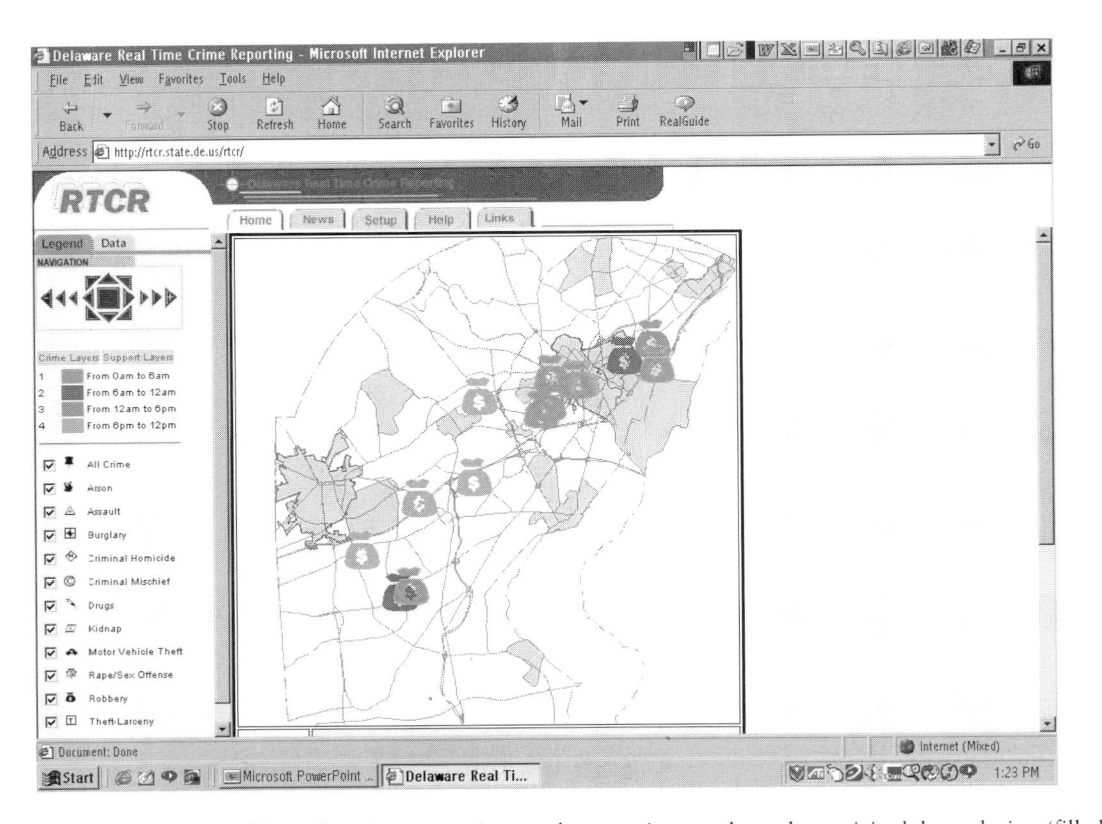

Figure 14.2 Regional map showing robbery locations superimposed on major roads and municipal boundaries (filled polygons) for Northern Delaware. Note customized symbol set. The moneybag is used to represent a robbery with color indicating type of robbery (in ESRI's symbol set a running man is used to portray robbery with various colors indicating various types of offense).

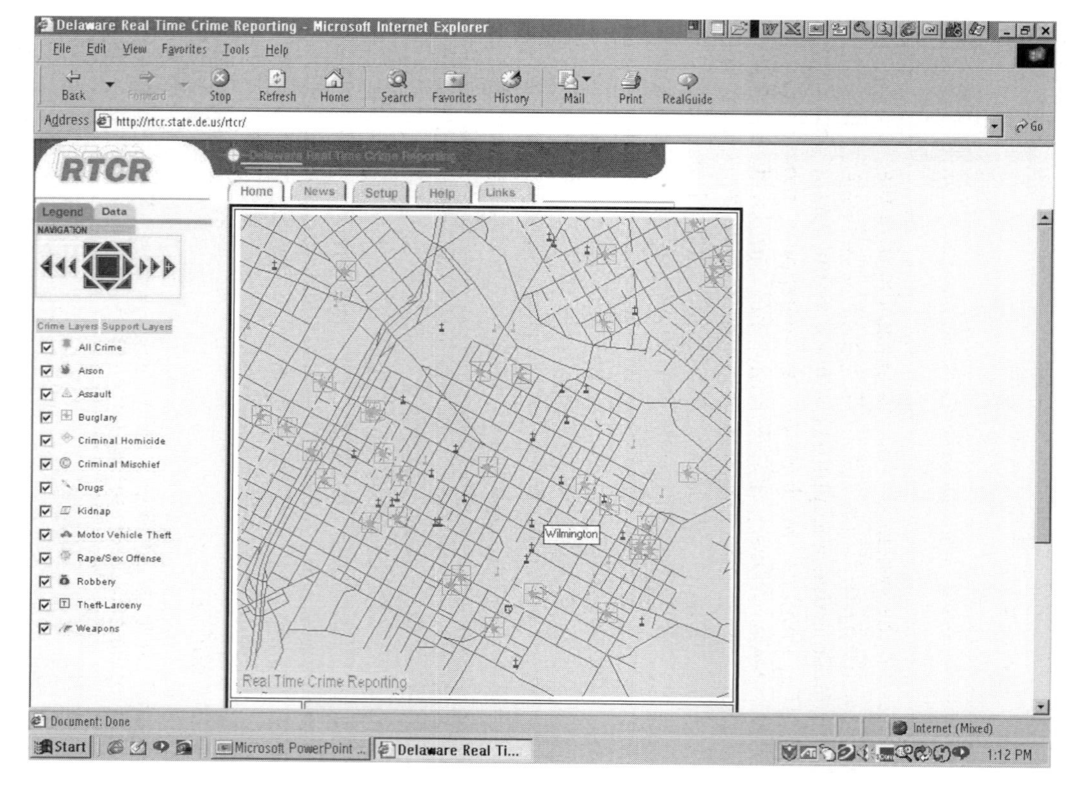

Figure 14.3 Large-scale view showing street centerlines within the jurisdictional boundary of Wilmington (Delaware's largest city). Note that various crimes are displayed. The map can be interactively zoomed from a statewide view to a detailed view with the navigator tool. The tool can also pan. Given the relatively small extent of Delaware this approach is intuitive and easy. For a GIS covering a larger area other tools such as a gazetteer function may be helpful.

Software is needed for application development, web functionality, and mapping. The following software is used:

- Microsoft Visual Basic 6.0 Enterprise Edition
- Oracle 8i with the Spatial Cartridge Option
- Intergraph GeoMedia Web Map and Web Map Enterprise and Microsoft Visual InterDev 6.0 were used for web development. The map output is active CGM or .jpg format.

Database design

Because the focus of the system is crime tracking, the data that is entered comes from crime reports. There are twelve types of crime that go into RTCR. The crime data that is captured includes: arson, assault, burglary, criminal homicide, criminal mischief, drugs, kidnapping, motor vehicle theft, rape/sex offense, robbery, theft/larceny, and any incident that is domestic violence related. The system tracks criminal incident data only; no calls for service are included. Incident data fields include: Delaware statute violation, date and time (occurred from and to), location (county, grid, sector, address, city, latitude, longitude, development, linear reference, and location type), reporting officer name and agency, whether it is domestic violence or hate crime-related, whether a vehicle, weapon, computer, alcohol, or drugs are involved, whether there is a suspect, complaint status, and whether report has been approved by a supervisor. Geographical features portrayed in the system includes street centerlines, municipal boundaries, churches, schools, parks, hospitals, police stations, and other landmarks. The officer in the field collects the majority of the critical data; that is the incident data. Some of the administrative fields, such as latitude and longitude coordinates and approval status, are derived within the system. The roadways data was purchased from GDT and the remaining geographical data was obtained from a number of local and state agencies.

Security, access, and accuracy

As the system is housed on a secure Intranet rather than the public Internet, privacy issues are minimal because the target users for the system are law enforcement staff only. All RTCR users must already have gone through the process of acquiring a DELJIS account, which requires fingerprinting and background checks. In addition, because the crime data is public record, there are no privacy issues related to release of information across jurisdictions.

The primary accuracy problem, as with most GIS, is the automatic geo-coding hit rate. In Delaware, because there was not a geo-coded statewide street centerline file, DPS was using various base-maps from the different cities and counties - some being more accurate than others. When the system went online, the automatic geo-coding hit rate was only 70–75 percent, but as of June 2000, it had been improved to about 85 percent in the most populous counties. DPS realized this was less than desirable and put in a system to handle the records that were not geo-coding automatically. As the counties develop and update their base-maps, statewide mapping accuracy will also improve.

Costs

The total development costs for the RTCR system were $732,287. This included hardware, software, licensing, and development costs. The costs were broken down as follows:

Software	$245,837
Hardware	$96,145
Integration/Installation	$266,150
Training	$74,125
Warranty	$65,130

Like most custom-developed technology applications, the majority of the costs were for the software, integration, and installation. Because this is a statewide system, and needed to integrate data from a number of agencies, these costs were higher than would be expected in a single agency implementing a crime-mapping application. The costs listed above do not include the individual local agencies' costs to implement their mobile systems. Substantial funding was made available by the state to assist the local agencies in going online with RTCR. Each agency submitted to the State its funding needs for equipment and resources, and this varied greatly according to size and existing resources. Agencies also paid their own costs of interfacing the RTCR system with other current technology efforts.

RTCR workflow

RTCR has two interface modes. The first is called QuickMap and it presents the user with a state map where icons denote the locations of crimes that have been reported in the last 24 hours. A different icon represents each of the twelve crimes now tracked. By clicking on the icon, the user calls up a text box containing details of the incident. The user can set default commands for the QuickMap display so it immediately targets the user's city or county and covers a time period preferred by the user.

To perform crime analysis, the user can click on what are called "advanced tabs." These allow the user to adjust the parameters of what will be displayed on the map. For instance, the map can present only certain crimes or crimes committed during a specific time or date range. The system can even draw a radial buffer around specific landmarks such as a school or shopping center to find out how many crimes are committed within a certain distance of it. If the user doesn't want to look at a map, query results can be viewed as a tabular list of addresses. Trends become evident when crimes are analyzed this way. DSP and local police officers feel that RTCR is helping them spot these trends sooner so they can take a proactive approach to stopping crimes before they occur.

RTCR features

RTCR has a number of useful and user-friendly features. These features include:

- Less than 12 hour turnaround time for crime data available for analysis
- Different mapping icons for each of the twelve crime types
- User-friendly query tool by crime type, date range, time of day, and felonies within a certain distance of a school or landmark
- Capability of adding other types of data as needed
- Data download utility for agencies to analyze their own data
- Ability to pan, zoom, and view details about an incident
- Graphic display of crime point maps and tabular summary listing
- Ability to identify exact location of crime.

Roll-out

Not only was Delaware the first state to enter the Union, but it also became the first state to implement a statewide crime-mapping system on 1 January 2000. The real-time aspect of the system's name reflects the rapid (no more than 12 hours) entry and geo-coding of data collected and submitted to the system. This up-to-the-half day (if not to the minute) data is a key aspect of the system's effectiveness and utility. According to Dover Police Chief Keith Faulkner (Retired) "In the past, by the time we received this kind of information, it had become history, and that's not good enough for law enforcement; we need to react quickly and be proactive. This RTCR system lets us do that."

Putting RTCR online

Prior to RTCR, Delaware already had established a statewide Intranet with T-1 lines for state agencies to communicate. Public Safety saw the Intranet as a practical and cost-effective means of making the crime reporting system available. Publishing GIS-based systems over an Intranet or the Internet is a relatively new technology. Fortunately, Intergraph was among the first companies to introduce a product, GeoMedia Web Map, specifically for this activity. Delaware purchased it and EIS implemented it.

"GeoMedia Web Map puts the RTCR on the Delaware intranet where any police agency has full access to its map viewing, data querying and crime analysis capabilities," according to EIS's Swanson. "With approved access, any department on the T-1 intranet can get on the system using either a Microsoft or Netscape web browser." The system is accessed most often by standard personal computers in each police station. Technically, RTCR can also be queried through one of the new MDCs in a patrol car, but they communicate via wireless cellular transmissions. The bandwidth is currently too small to make direct RTCR access practical, but this will undoubtably change as telecommunications systems are upgraded in the future.

Examples of applications

Users of the RTCR system are found in all types of agencies, from the Governor's Office to the State Department of Public Safety, from little hamlets like Rehoboth Beach to major cities like Wilmington. Patrol officers, commanders, analysts, and investigators are all using the RTCR system to meet a variety of needs through its many functions. Commanders can determine "What type of crime occurred within the last 24 hours? At what time and locations are these crimes occurring? Are there similar crimes in the area?" RTCR allows users throughout Delaware to know within hours what type of crime is happening and where. The system is also being used to hold Police Chiefs and troop commanders accountable for what is going on in their jurisdictions. This tool can show crime data and trends comparable across areas or time.

In the short time that RTCR has been online, it has been used primarily in the precincts and squad rooms where the sergeants are preparing the officers for their shifts, according to Colonel Gerald Pepper of the Delaware State Police. "The sergeant comes to work prior to the shift meeting, accesses the map of his part of the county and looks at what crimes have occurred in the past 12 hours, one week or one month," Pepper said. Continuing, he noted "From that, he or she will develop hot spots and know where to deploy the troopers."

RTCR has been used to assist arson investigations, identify burglary trends, stay current with bordering jurisdictions' criminal activity, and adjust patrol strategies during the beginning of shift roll-call meetings. As of May 2000, DSP had deployed 300 mobile data terminal units in addition to those MDT units belonging to the forty-two local agencies.

Mary Ann Papilli, a DSP Captain, noted a number of uses of the RTCR. "Troopers can now access current crime data, shift sergeants can monitor the status of reports as well as view recent crime on their shift in their sector, and administrators can do simple maps and analysis to look at trends and patterns." Troopers, from patrol and investigations, have used RTCR to make maps to bring to community meetings to inform the public about a series or trend. Lieutenants and captains are finding RTCR to be a great administrative and information sharing tool. Not only are they using it to export the data into other software for further analysis, but also Captain Papilli has recently used the system to examine neighboring jurisdictions to see if her pattern of a rash of burglaries extended outside of her troop's area.

Chief Faulkner, formerly from Dover Police Department, also refers to the RTCR as a great tool. He personally used it to monitor trends and encourages others to use the system and its data and maps to bring problems to his attention. DPD generates monthly statistical analyses and maps and posts these in the roll-call room as well as providing them to staff and road supervisors. The Chief recently learned about an auto theft problem from one of these reports, and was able to knowledgeably reallocate personnel to work on the problem. In addition, his investigators are constantly using the RTCR to identify similar cases in other cities in the county as well as the state so as not "to be isolated by political boundaries." Chief Faulkner admits that DPD and other departments are not yet maximizing the use of the RTCR, but he is strongly encouraged and optimistic by what he has seen so far and where they are going with it. Instead of relying on the old paper environment of writing a report, getting it approved, sending it through the proper channels and data entry avenues before being able to review the information in any helpful format, officers in the field now have a very up to date, useful, and practical tool for extracting data. This application and the data are also easily distributed through a web browser interface, cutting down on technical training and upgrade and maintenance issues for a distributed system.

RTCR is an additional weapon in the arsenal against crime in Delaware. Officers and troopers now use data to drive their decisions. Instead of being traditionally reactive, Delaware law enforcement uses the system to take an analytical approach to crime control and crime prevention. The application helps to identify problem areas, focus police resources, and support strategy development to address public safety issues. Commanders use the information to identify crime trends and deploy their officers to hot spots in hopes of reducing crime in their area. Patrol officers and supervisors use the system to direct their patrols and help to apprehend offenders.

Crime control and prevention have been found to be more successful when data is shared – between units within an agency, across agencies, and with the community. The foundation of RTCR is ideal for interagency sharing of incident data. As noted in the above examples from the DSP Captain and Dover Chief, crime series and patterns can be better identified and analyzed when looking at information that crosses jurisdictional boundaries. All too often, a void in the data shared between separate agencies becomes an obstacle to effective crime fighting and public safety. The statewide crime tracking and mapping system is bridging the information gap.

Inter-agency cooperation

Delaware State Police have also been innovative in working with municipal law enforcement agencies in addressing local law enforcement problems that may spill over the boundaries of a single jurisdiction or may require resources beyond that available to a beleaguered department. This effort has used GIS extensively and has been in place prior to the rollout of the RTCR system. An example of interagency cooperation facilitated by GIS-based crime analysis was cooperation between the DSP and Wilmington, Delaware Police Department to deal with an upsurge of violence related to drug dealing, particularly of crack cocaine.

Challenges

Implementing any new technology across jurisdictions and disparate users is always a challenge. Some departments had personnel that were comfortable with computers and quickly adjusted to the new system, while others are still attempting to get their compatible systems online. The latter has been the greater challenge – getting every agency on board. The only incidents that are sent to RTCR are those where EPC, the field reporting system, is used.

The amount of user involvement at the onset of the RTCR project has minimized major implementation problems. For the most part, the system has evolved into exactly what was originally envisioned. Because the user interface is straightforward and easy to learn, technical training has not been an issue. The larger training issue, which is often overlooked in many projects but is being gradually done throughout the Delaware agencies, is how to apply this new tool. Continuous training is available to inform troopers and officers at all levels and responsibilities of the full functionality and usefulness of RTCR.

Another challenge, discussed earlier in the "accuracy" section, is the automatic geo-coding hit rate. For instance, one county is still working on their 911 unique addressing project. This has hindered address-matching capabilities there. Although DPS has overcome the lower geo-coding rate with a manual coding system, this has delayed getting the accurate data into the system and available for use. This and other "system management" issues are being ironed out as they arise.

Conclusion

Although Delaware is unique in the United States in terms of the responsibilities of the State Police, it has partially therefore chosen to use GIS as a common statewide mapping, reporting, and analysis approach to ease the law enforcement burden for both its own officers and those of the other agencies in the state. Such comprehensive interagency cooperation is the exception rather than the rule in the United States but, using GIS as a unifying technology, perhaps other states can begin to develop statewide crime mapping and reporting capabilities of their own.

15 National Guard Bureau – Counterdrug GIS Programs

Supporting counterdrug law enforcement

Billy Asbell

Use of GIS and other geo-spatial technologies at the National level is not unknown around the world, but it is rare. This is perhaps because those nations with very powerful and intrusive national police forces are generally not the free market democracies whose prosperity has engendered the wherewithal to obtain and use GIS. Exceptions to this rule might include Germany, Singapore and the Czech Republic. Elsewhere, National Police agencies are less likely to be active users of GIS than are local and regional law enforcement entities. In the United States the Federal agency that uses GIS most actively is probably the US Border Patrol, which employs GIS, false color infrared digital imagery, GPS and a whole range of geo-spatial technologies (many deployed in the field) to help interdict the massive flow of illegal immigrants, largely across the US southern border with Mexico. Another example of use of a wide range of geo-spatial technologies by an agency of the US Federal Government is provided by the Counter Drug Office of the National Guard Bureau. They use GIS, spatial decision support systems, aerial imagery, digital terrain models, classified aerial photography, GPS and real time transmission of video imagery in efforts to assist the interdiction of drugs and the detection of illicit drug laboratories and marijuana fields.

(Editor)

The United States National Guard Counterdrug Program (NGB-CD) is a national counter-narcotics program utilizing the National Guard membership in all of the 50 states. This program provides highly skilled personnel, specialized equipment, and facilities to support counterdrug law enforcement agencies (DLEAs) and community-based organizations to better respond to the changing drug threat in our Nation. It is our intention to be recognized by law enforcement entities, community-based organizations and our citizens as the premier support force seeking to create a drug-free Nation. Since 1989 the National Guard has been supporting law enforcement organizations in their efforts to counter the drug threat in the United States. On any given day the National Guard Counterdrug Program has over 3,000 Guard personnel on active duty nationwide supporting all the fifty states and four territories with counterdrug programs. These efforts include providing personnel, training, facilities, assets, knowledge, methodology, and access to emerging technologies such as geographic information systems (GIS), remote sensing image processing, and global positioning systems (GPS).

GIS initiatives overview

The objective of NGB-CD's GIS initiatives is to integrate existing technologies and counterdrug assets in support of the National Guard Bureau Counterdrug Program's (NGB-CD) mission to eliminate the drug threat in the United States. The NGB-CD recognizes that marijuana will remain the most commonly abused illicit drug in the United States and that technology holds great promise for strengthening the capabilities of law enforcement agencies in identifying and eliminating cultivation and distribution of marijuana.

These initiatives take the form of activities in two main categories: operations and training. In the area of operations, assessment of available data is giving rise to the counter drug geographical assessment regional sensor system (CD-GRASS) program. To address the unique training requirements of geographical information systems (GIS), the counterdrug geographic information systems training (CD-GIST) program has been developed. These two programs, working in concert, will provide greatly enhanced capabilities to law enforcement. The following sections describe some of the initiatives and activities we support.

DMC (digital mapping center)

The digital mapping center (DMC, formerly known as the digital mapping initiative, DMI) has been in operation since 1992 providing DLEAs with free computer-generated mapping products custom designed for their needs and applied to everything from mission planning to intelligence analysis. Maps are produced from digitized data provided from numerous sources. Map products can be produced in a variety of scales from 1:2,000,000 to 1:24,000 or larger for customers with local interests (Figure 15.1 shows a map view of Nogales, AZ) to National interests (Figure 15.2 shows an overview of the High Intensity Drug Trafficking Areas (HIDTAs) produced for the Office of National Drug Control Policy (ONDCP)). Products can be provided showing terrain features, aeronautical information, rivers, lakes, counties, cities, roads, highways, latitude–longitude grids, towers, and more. In addition, maps can be customized to include arrest statistics, seizures, marijuana growing statistics, etc. Location data (either GPS or latitude–longitude information) for customized maps must be provided by the requesting client, who can obtain the coordinates from many sources including autonomous handheld GPS units. Street-level mapping products derived from US Census Bureau TIGER data can also be provided at scales

Figure 15.1 The Digital Mapping Center developed this terrain model with draped aerial imagery to help drug enforcement and border patrol agents interdict drug traffickers utilizing the rough terrain of the US Mexican border to bring drugs into the US in the vicinity of Nogales, Arizona. Image supplied from National Guard Bureau Counter Drug Office.

of 1:12,000 to 1:10,000. All products are provided free of charge for use in a counter-drug mission. In fiscal year 2000, the DMC provided fifty thousand pages of maps to clients. The largest group of users is state and local law enforcement agencies, which requested fifty percent of the DMC's production.

CD-GRASS (counterdrug geographical regional assessment sensor system)

The NGB-CD, through the CD-GRASS program, led by the Georgia Tech Research Institute, is incorporating an innovative blend of GIS, digital mapping, enhanced visualization, virtual reality, electro-optics, sensor fusion, modeling and simulation, data acquisition, and other operational assessment technologies. The mission of CD-GRASS is information integration, assessment, and decision support to counterdrug operations through state-of-the-art technology transfer. The CD-GRASS program also incorporates National Guard Bureau assets including the DMC, aerial reconnaissance and sensor deployment, law enforcement training centers, and the fifty-four state and territory counterdrug task forces.

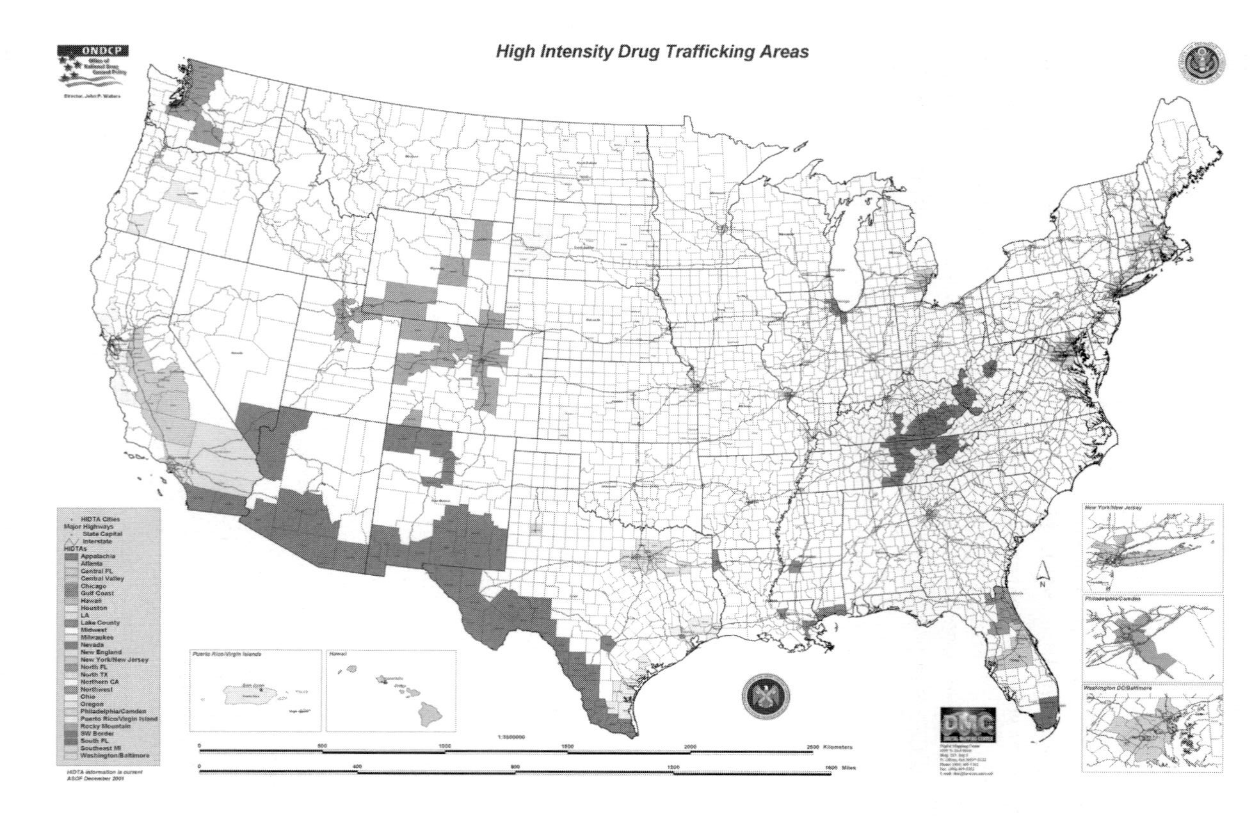

Figure 15.2 This Nationwide map showing High Intensity Drug Trafficking Areas (HIDTA) Map was produced for the National Guard Bureau Counter Drug Office by the digital mapping center (DMC). Darkest counties have the highest intensity of illicit drug trafficking and/or cultivation.

The National Guard Bureau conducts numerous aerial reconnaissance missions using aircraft such as twin-engine turbo-prop aircraft and helicopters. These aircraft serve as platforms for a wide range of sensors, which generate a variety of imagery and locational data that is processed and synthesized for spatial analysis using GIS.

Every mission generates some level of information whether it is data, intelligence, non-imaging sensor data, evidence, observations, notes, mission location use of GPS, maps, and remote sensing data (such as that generated by Forward Looking Infrared (FLIR), digital cameras, and multi/hyperspectral imagers). This information needs to be analyzed and integrated in some manner to insure effective operations.

Information assessment is the review and analysis of collected information. The goal of such assessment is elucidation of patterns and relationships leading to improved operations. This can be accomplished through mapping/GIS analysis, spatial/spectral/temporal modeling of potential targets, target detection and prediction, or a decision support system (DSS) as shown in Figure 15.3. The outcome of this analysis can then be used to focus future law enforcement operations, such as raids on illicit drug growing operations.

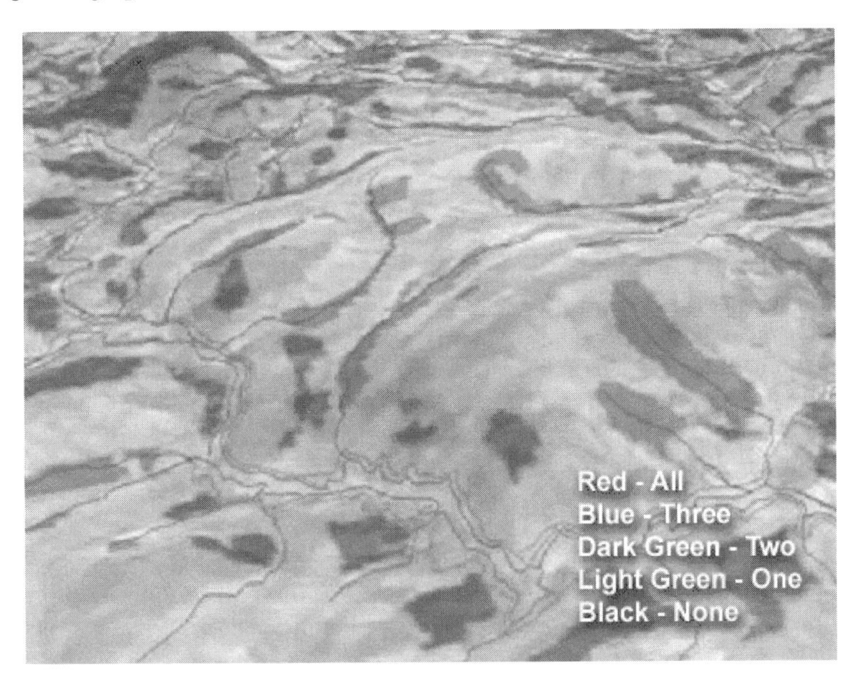

Figure 15.3 Aerial imagery is draped across a terrain model and color coded to indicate areas for higher probability of containing illicit marijuana fields in this criteria-based sample map product. (This was created using a model developed by the Georgia Tech Research Institute for the National Guard Bureau Counter Drug Office.)

Operations are activities based on information assessment resulting in new intelligence and mission strategies. The use of automated GPS/geo-coded mission information, aerial/ground coordination and communication, virtual GIS, and mission planning and evaluation can greatly aid local and state law enforcement in conducting raids and in destruction of drug crops.

DSS (decision support system – MCEDSS)

As part of the CD-GRASS program, the NGB-CD has also funded the University of Southern Mississippi's Center for Higher Learning (USM-CHL) to design and build a Decision Support System (DSS). The goal of this DSS is to develop a system that will increase the efficiency of the planning and conduct of marijuana eradication operations. The ultimate users of the DSS will be the State DLEAs, and the National Guard Counterdrug personnel in support of counterdrug operations. One of the fundamental problems relating to the eradication of marijuana cultivated outdoors is the size of the search area relative to the available resources. The 'heart' of the DSS is a rule-based expert system module that identifies areas with a high potential for marijuana cultivation as shown in Figures 15.4–15.6. These three successive figures illustrate a simplified approach of the DSS. Figure 15.4 shows the radial proximity boundaries of offender locations; Figure 15.5 shows another proximity boundary around roads and streams; and Figure 15.6 shows the summary of all of these layers indicating high probability areas for marijuana cultivation.

The DSS will have a single headquarters node consisting of a cluster of servers and workstations that access and process data to produce a suite of map layers. The primary map layer will provide verification of geographic areas with the highest likelihood of marijuana cultivation sites. This layer will be the output of a rules-based expert system. The decision layer, supplemental map layers, and other information will then be distributed to law enforcement agencies via an Internet-based communications network. Field nodes (aircraft and ground-crew laptop computers) will also be able to download this information at a district office via the network for use during their search/eradication operations in the field.

The DSS is designed to take maximum advantage of the DLEA's existing hardware and communications systems, and concentrates new hardware and software components in a single headquarters node to reduce implementation costs. Additional cost savings are achieved through the maximum possible use of commercial off-the-shelf components.

DART (database for assessment of requirements and tactics)

Prior to the deployment of the DSS and other systems from the CD-GRASS program, a clear understanding of the current law enforcement agency information technology infrastructure, interface, and architecture must be

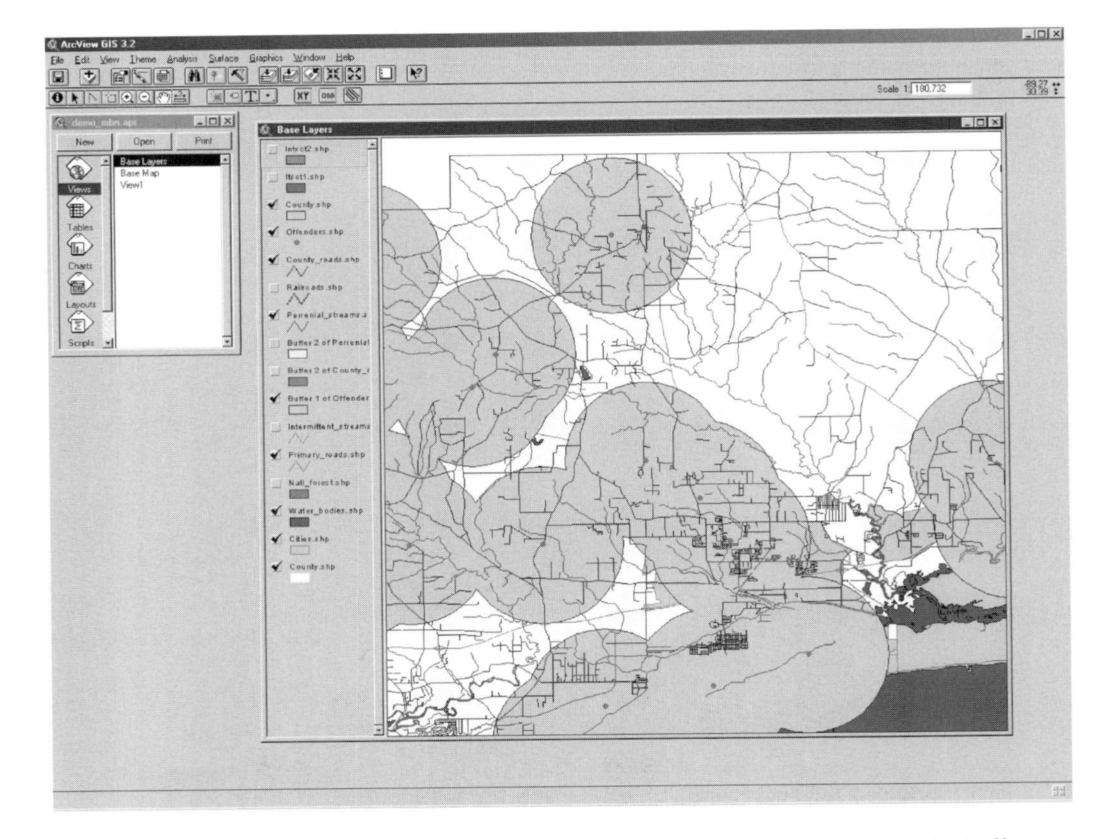

Figure 15.4 A spatial decision support system (DSS) was used to define search areas for illicit drugs. The buffer zones are a first step in an iterative process of identifying search areas. (The work was performed by the Center for Higher Learning at Stennis Space Center on behalf of the National Guard Bureau Counter Drug Office.)

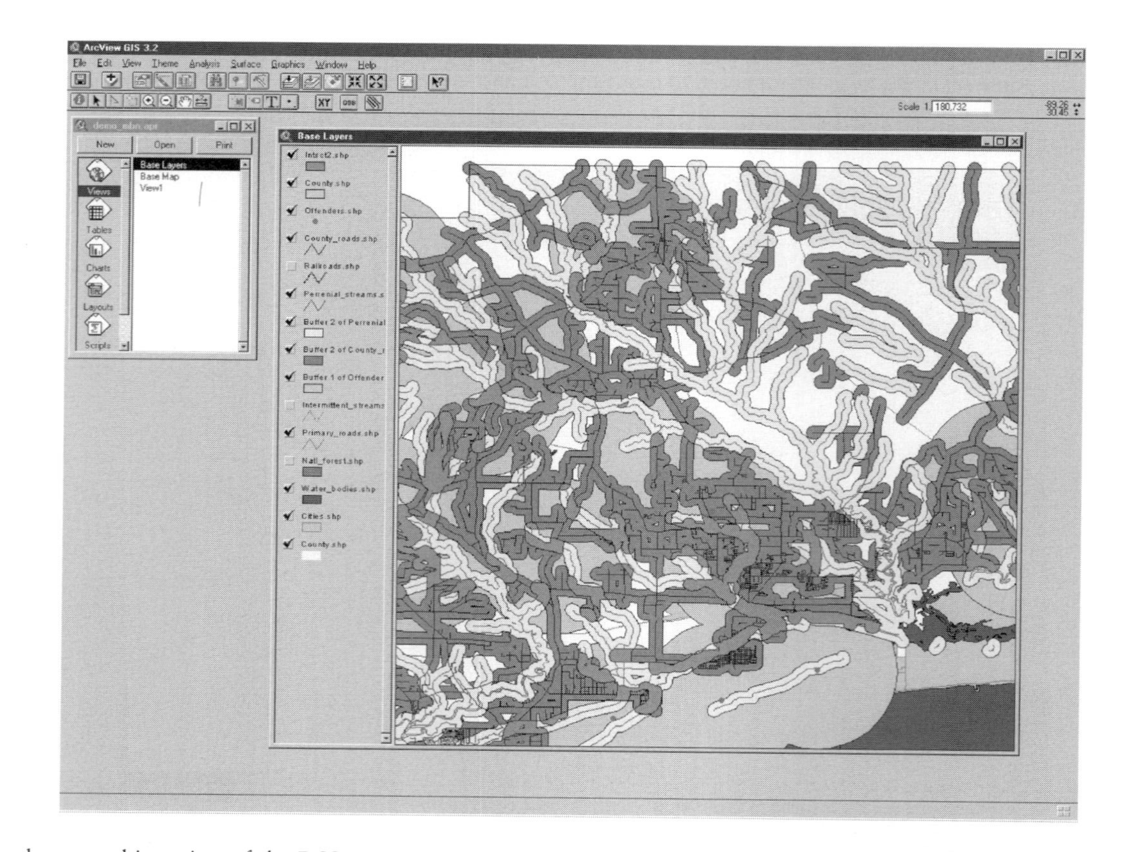

Figure 15.5 In the second iteration of the DSS process; search areas are narrowed down based on environmental factors.

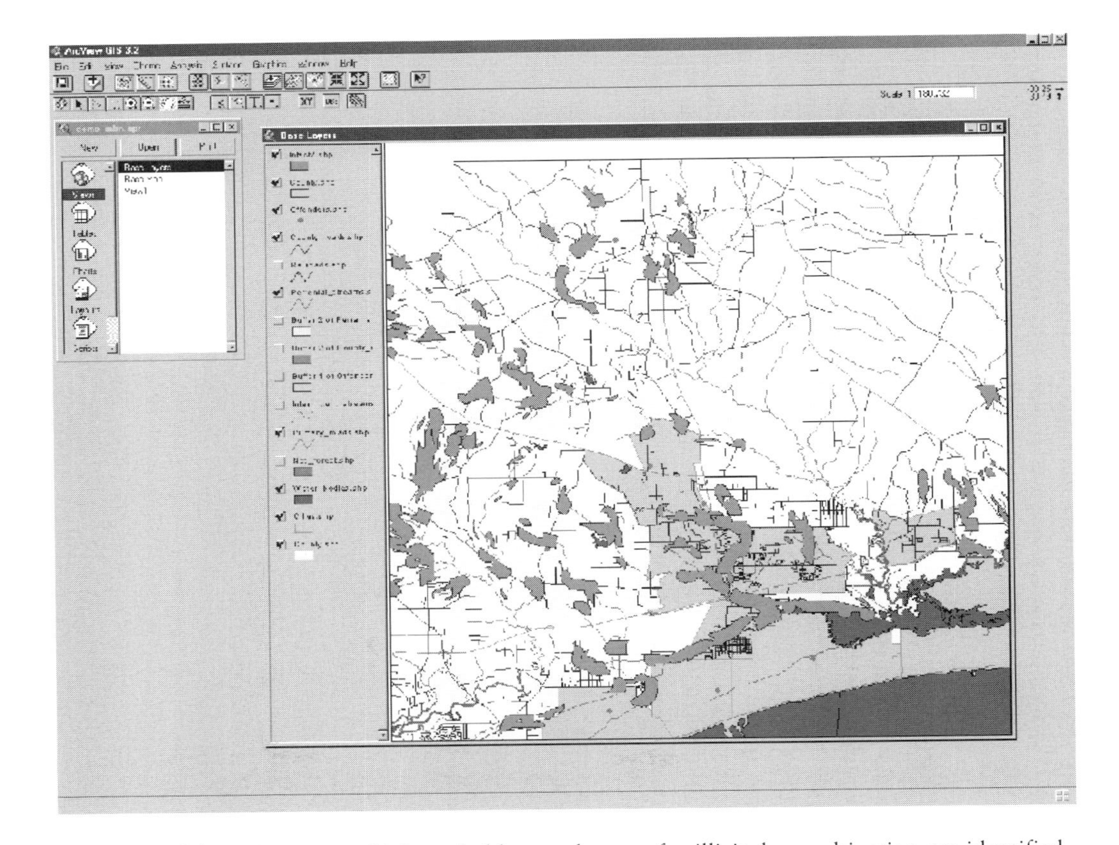

Figure 15.6 In the third step of the DSS process; high probably search areas for illicit drug cultivation are identified.

documented nationwide. This task requires a GIS interface to insure localities are documented and spatially represented. To accomplish this task the database for assessment of requirements and tactics (DART) has been developed. This tool is envisioned to greatly enhance the planning capabilities of the National Guard Counterdrug Coordinators at the state level in their support to DLEAs. It will provide an electronic capability, using GIS, to quickly visualize opportunities to reduce asset deployment while enhancing the DLEAs operational impact. It will also serve as a standardized contact database for DLEAs (Figure 15.7).

DMS (digital mapping server)

To address the extensive spatial land attribute data requirements to support a GIS environment, the digital mapping server (DMS) project has also been initiated. In collaboration with the Federal Geographic Data Committee, this project has been established to ensure conformity of all products to the National Spatial Data Infrastructure standards. This technology is based on current commercial and Government initiatives underway such as the Geography Network, the Open GIS Consortium, and Digital Earth. This system will allow all levels of government to access GIS data from verifiable

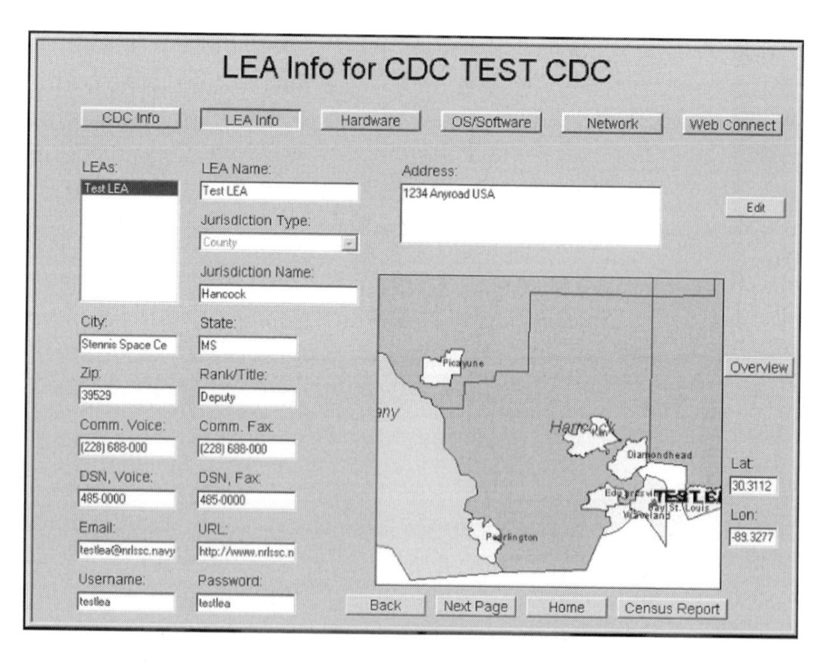

Figure 15.7 This image shows the mapping component of the data entry screen for the Database for Assessment of Requirements and Tactics (DART). (This system was developed by the US government at the Naval Research Lab – at the Stennis Space Center in Gulfport Mississippi.)

sources via the next generation of Internet by simply querying DMS. This will be the GIS data source for the CD-GRASS program.

Hyperspectral/Spatial

The NGB-CD in partnership with the office of national drug control policy and the drug enforcement administration are working to develop hyperspectral signatures for illicit crops to support airborne and space-based detection. This technology is based on light absorption and reflection of plants. Currently the US Department of Agriculture – Agriculture Research Service, West Virginia University, and the Georgia Tech Research Institute are working on developing these signatures. While this activity shows great promise, it is not the only necessary component of an effective system. To assist in finding illegal crops on cloudy days or periods of low light, spatial analysis will also be employed. Using either of these technologies or both, illegal crop growth should be detectable based on the light signatures as well as the linear aspects of the fields or other characteristics of the growing process. These signatures will be overlaid onto the GIS product created by the DSS to further narrow the possible plot locations.

Counterdrug geographic information systems training

To address training requirements, the National Guard Bureau has begun the Counterdrug Geographical Information Systems Training Initiative. It focuses on training law enforcement professionals on related aspects of GIS principles and software. Through the efforts of the Federal Law Enforcement Training Center, the four National Guard Bureau schools and the National Guard Bureau Environmental Programs Division, law enforcement officers and staff will learn how GIS can assist them in their mission. Participants will also learn how to use the most common commercially available GIS software. With support from the crime mapping research center, this program is intended to change the way DLEAs maintain historical documentation and the methodology they use to conduct investigations and operations.

Technology evaluations

The National Guard Counterdrug Office is pursuing other technologies to support DLEAs that have some relation to GIS activities. One of the many services NGB-CD provides counterdrug law enforcement is the evaluation of currently available products that may be acquired to increase their capability as well as effectiveness. Part of this effort includes providing integration schemes such as Figure 15.8. The following examples are a sampling of systems we have evaluated that have a relationship to GIS and other geospatial technologies.

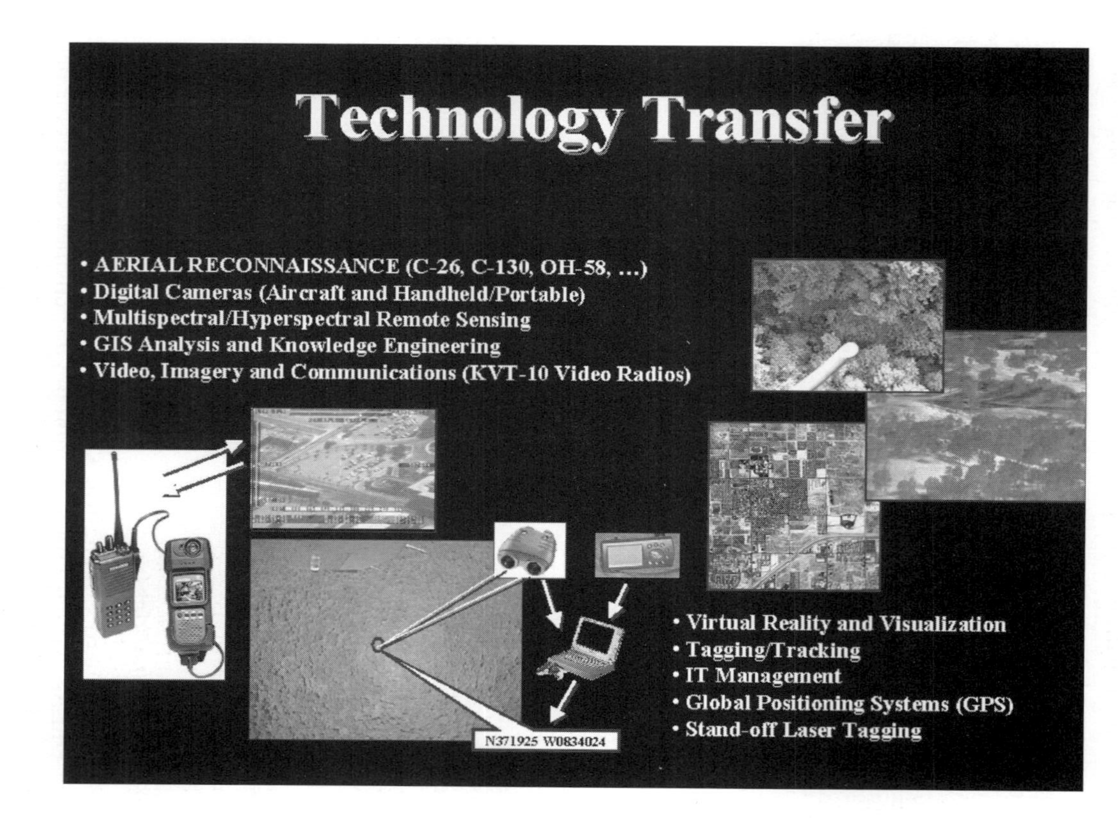

Figure 15.8 Integration of geospatial technologies by the National Guard Bureau.

GPS binoculars

The Leica, a German optical company, has developed several different models of GPS binoculars that have the capability to acquire the GPS coordinates for any position within the range of the laser range finding device in the binocular. NGB-CD tested a pair of these binoculars (Figure 15.9) that have a range of over 4,500 meters. They were built to work with a "PLUGER", a large military GPS unit mounted in a vehicle or aircraft, to acquire the GPS data information. The Georgia Tech Research Institute in conjunction with the Naval Research Laboratory – District of Columbia modified the software on this device to change the GPS support from a high dollar, highly complicated device to a less expensive Garmin GPS unit selling for less than $200. In addition, other modifications allowed the creation of a digital database on a laptop connected to the system resulting in the GPS coordinates for an actual site, verses the aircraft location, to be automatically recorded in a database utilizing a low cost GPS device. Thus, a helicopter can hover unnoticed away from a suspected drug lab or marijuana growing site but get the exact coordinates of the site itself stored in a program that can be downloaded to a base map in a GIS. This map can be used by law enforcement to navigate to the often isolated site hidden in a rural or forested area and stage a subsequent raid.

Kenwood KVT-10 Video Radio

The KVT-10 (Figure 15.10), another commercial product, allows the transmission of video still images across the 800–900 MHz band of a handheld radio or via a cell phone. The test allowed an aircraft to downlink its picture to a handheld wireless unit up to forty-five nautical miles away at an average speed of one color frame per thirty seconds. It allows for the transmission of digital video in both directions (air-to-ground and ground-to-air).

Figure 15.9 GPS binoculars developed by Leica, Incorporated and used to pinpoint illicit drug operations from helicopters hovering at some distance away from the targets of suspicion. GPS coordinates of the targets can be acquired without coming into dangerously close proximity. Field teams can later approach these targets on the ground with guidance from hand held GPS units (Image Courtesy Leica Corporation).

Figure 15.10 KVT-10, a COTS product.

The test also demonstrated the ability to download these images on the fly to a laptop attached to the system. These systems are available now and cost around six hundred dollars per unit. The KVT-10 can also be used with cellular phone technology. The KVT-10 was also able to download digital forward looking infra-red images off of the aircraft's sensor suite into ground-based laptops of law enforcement for use as backdrops for vector-based maps. This capability not only showed how to approach a suspected site but also highlighted features such as possible hiding places and escape routes.

Digital camera upgrade to KS-87

Our airborne reconnaissance assets are in transition from a wet film KS-87 camera to a digital capability. A six-month evaluation by the Georgia Tech Research Institute allowed NGB-CD to move from a 4-megapixal system to an 85-megapixel system for basically the same cost. This kept the upgrade from degrading the current wet film capability while utilizing our current camera bodies. The modification resulted in extending the useful life of a Vietnam war era reconnaissance camera. This digital imaging system will be geo-referenced with airborne GPS for inclusion as raster layers into GIS without the delays involved in printing, scanning, and registering the images encountered by pre-existing photogrammetric methods.

Collaboration

Teaming efforts are the key to success by reducing costs and increasing cross agency inter-operability. Evidence of the achievements in these efforts is demonstrated by the interaction with our partners. The following agencies are partners with the National Guard Bureau in this sponsored activity:

Georgia Tech Research Institute: The Georgia Tech Research Institute is leading the CD-GRASS program. Their responsibilities include technology transfer, operational insertion, technology evaluation, systems integration, and program coordination.

Center for Higher Learning: The Center for Higher Learning at Stennis Space Center has developed the Mississippi Counterdrug Enforcement Decision Support System (DSS), which will be integrated within CD-GRASS.

West Virginia University (WVU): In conjunction with the other university partners WVU is working on hyperspectral detection of Cannabis from aerial platforms using GIS as the underlying viewing and analysis tool.

Digital Mapping Center (DMC): A primary goal of CD-GRASS is to increase DMC's capabilities to support counterdrug law enforcement. DMC provides law enforcement agencies with free computer-generated mapping products annotated to their specifications for their operations from mission planning to intelligence analysis.

The Department of Justice/National Institute of Justice Crime Mapping Research Center (CMRC): CMRC focuses on aiding law enforcement in GIS efforts. Their partnership in the CD-GIST and DMS projects will enhance NGB-CD's capability to train law enforcement in the use of GIS systems as well as provide access to large amounts of GIS data. This association will further the effective employment of CD-GRASS while at the same time assisting the CMRC in their efforts to train law enforcement in GIS.

Federal Law Enforcement Training Center (FLETC): FLETC trains most of the State and local DLEAs as well as a large number of the Federal officers. They are working with NGB-CD to establish a GIS basic course within their core curriculum. This course will insure the DLEA community receives a basic understanding of GIS and how it can be exploited.

Counterdrug Technology Development Program: The Counterdrug Technology Development Program has been instrumental in assisting NGB-CD with the acquisition of technology for counterdrug field operations.

Naval Research Laboratory – Stennis Space Center: The Naval Research Laboratory at Stennis Space Center has developed a Database for Assessment of Requirements and Tactics (DART) and provides expertise in advanced mapping and GIS technologies. This program will assist in determining the level of information technology available nationwide within the law enforcement community in preparation for the deployment of CD-GRASS.

Naval Research Laboratory – Washington, DC: The Naval Research Laboratory in Washington, DC is developing the necessary components to deploy laser range-finding, remote GPS unit binoculars for field use in automated mission tracking and to improve ground-level positioning accuracy.

US Department of Agriculture – Agriculture Research Service: The Agricultural Research Service is analyzing remote sensing techniques required for spotting marijuana via its spectral signatures. They are also building growth and yield models for illicit crops. These two components will be leveraged within CD-GRASS.

US Forest Service: The US Forest Service is supporting the CD-GRASS program by loaning imaging equipment, providing field expertise, coordinating aerial platforms, leveraging specialized GIS information and assisting in operational exercises. In addition, the Southern California Regional Office has been involved in a non-automated proof of concept of the basic elements of CD-GRASS.

Carolinas Institute of Community Policing: The Carolinas Institute for Community Policing is a training partnership of the Charlotte-Mecklenburg Police Department; US Department of Justice, Office of Community Oriented Policing Services; Pfeiffer University at Charlotte; the University of South Carolina at Columbia; the University of North Carolina at Charlotte; and South Carolina Educational Television. The institute has trained over 1,600 police, sheriff, and community members in topics such as Introduction to Community Policing, and Community Policing Problem Solving and Advanced Problem Solving. The Institute is furthering NGB-CD support to DLEAs by providing export locations and instructional assistance in the area of GIS technical training.

Crime Mapping and Analysis Program: The Crime Mapping and Analysis Program provides technical assistance and training to state and local agencies in crime and intelligence analysis and geographic information systems. The Program is working in the NGB-CD collaborative effort to insure that GIS instruction maintains a level of common core competency providing a clear asset for law enforcement agencies.

National Guard Bureau-Army Environmental: The Environmental Programs Division within the National Guard Bureau is engaged in supporting common GIS technologies and education. These efforts include the CD-GIST and DMS initiatives. Their contribution focuses on assisting in GIS training and collaborating on the GIS data server issues.

Drug Enforcement Administration: A strong supporter of our efforts. The Drug Enforcement Administration is working with other agencies to assist in the defining of a hyperspectral signature for cannabis.

Office of National Drug Control Policy: Via the Counterdrug Technology Assessment Center several efforts are being pursued. These include the development of hyperspectral signatures and GIS technology.

National Guard Bureau Schools: Schools located in California, Mississippi, Florida, and Pennsylvania play a key role in instructing the DLEAs on GIS software.

State National Guard Units: Once deployed, CD-GRASS will be implemented in all fifty states and four territories.

Conclusion

The objective of the National Guard Bureau's Counterdrug Office's GIS initiatives is integration of existing technologies and counterdrug assets in support of law enforcement's counterdrug mission. This objective is being attained through the integration of the sub-elements of the counter drug geographical assessment sensor system (CD-GRASS) program. The survey information gained from the DART system will be used to determine the deployment plan for the decision support system (DSS). The DSS will incorporate law enforcement agency data with the GIS data served up by the digital mapping center to create a product for counterdrug law enforcement. This product will predict cannabis growing sites locations with a high probability based on knowledge-based expert system and historical law enforcement data. The Digital Mapping Center will continue to support requirements of counterdrug law enforcement with custom generated cartographic products until all users have access to the CD-GRASS program. Remote sensor data layers, such as hyperspectral and spatial signature information, can then be included in the GIS, finetuning the predictions. The use of the products from the technology evaluations will enhance the detection and eradication operations, further increasing the effectiveness of the overall program. The use of GIS software and by-products, will become more widely accepted through the free GIS training offered to participating agencies. Finally, a consortium of all users will be encouraged to collaborate on development of the evolving approach. Such cooperation will reduce costs, increase inter-operatability and enhance the effectiveness of GIS and geo-spatial technologies in combating illicit drugs. The National Guard Counterdrug Program is dedicated to insuring not only the success of internal National Guard Programs but also the success of associated law enforcement efforts to interdict illegal drugs and apprehend drug growers, manufacturers, and traffickers.

16 Journeys to crime

GIS analysis of offender and victim journeys in Sheffield, England[1]

Andrew Costello and Mark R. Leipnik

GIS is being used almost everywhere for almost everything throughout the world, from cartographers in Mongolia and petroleum geologists in the Sudan, to market researchers in Manhattan and sociologists in the green and pleasant precincts of England. In this chapter is presented a spatial crime analysis application from England. England and the United Kindom generally, is a world leader in socio-economic applications of GIS. The tremendous advantage derived from the conversion of the highly detailed and generally quite current Ordinance Survey maps of the United Kindom into a GIS format, finds its expression in applications like the one presented below. Few countries can match the detail, accuracy and sophistication exhibited in law enforcement applications of GIS in Great Britain, since few countries (outside of Scandanavia and Switzerland) can match the detail and accuracy of the geo-spatial data that is available in places like Sheffield, England.

(Editor)

This paper reports on a GIS-based analysis of patterns of "travel to crime" in the city of Sheffield, England. As well as identifying the patterns of offender travel there was also an examination of how the location and type of victimization of residents of a number of areas varied. The identification of offender movements into and within an area is obviously important in terms of developing crime prevention strategies. We would like to be able to determine whether we are dealing with 'insiders' or 'outsiders'? By the same token we could argue that understanding how and where the residents of an area come to be victimized is vital in targeting crime prevention strategies.

Data

The recorded crime data for Sheffield covered calendar year 1995[2] and was provided by the South Yorkshire Police. This amounted to linked details of over 70,000 offences, involving 54,000 complainants (victims) and 8,000 known or suspected offenders. The three datasets were geo-coded to enable us to link them spatially to a base-map of every residence which in England has a unique postal code; the location of an offence, the home address of any victims and, if known, the home address of responsible offenders. The geo-coded data was supplemented by a series of interviews with local burglars and car thieves to both test the geo-coded data and add explanatory value to the analysis results.

The journey to crime

Traditionally, research into the 'journey to crime' has looked at the travel undertaken by offenders, from their nominal home address to the place the offence was committed (see, e.g. Pyle 1974). This is obviously an important field of study, but potentially ignores a major element of journeys to crime, that of the victim or offence target[3] to the place of offence commission. It could be argued that in order to understand the patterns of crime in any given area/place one needs to look at not only how offenders come to be in a place/area but also the travel patterns of their victims or targets. In a similar vein, it can be argued that to understand the patterns of victimization of residents of any given area one needs to know both about the journeys of offenders and victims.

The journey to offend

The following findings emerged from the analysis of the journey to commit offences data:

- The majority of journeys involved in offending are relatively short – over 40 percent of burglary journeys were less than one mile.
- Interviews suggested that travel associated with crime was not primarily driven by plans to offend, but offences occurred as opportunities presented themselves during day to day routines.
- When journeys to offend are undertaken they tend to be short and longer-range travel tends to be to places the offender has links to.
- When compared to previous research we found little evidence of increasing journeys or that new travel opportunities were changing offender travel patterns.

The journey to victimization

The journey to victimization varied by the type of neighborhood the victim resided in. In wealthy neighborhoods up to half of victimizations occurred well away from where the victim lived. In contrast, in poor neighborhoods up to 90 percent occurred within a few hundred yards of the victim's home. Table 16.1 shows the differences in travel for two neighborhoods: one with a high victimization rate, high offence rate, and resident offender rate, the

Table 16.1 The journey to offend and the journey to victimization in two contrasting Sheffield suburbs[4]

Area crime type	*Average journey to offend in area (miles)*	*Average distance from home when victimized (miles)*	*Proportion of victimizations within neighborhood (in %)*
Area 1: High offender/offence/ Victimization[5]	3.51	1.73	55
Area 2: Low offender/offence/ Victimization[6]	0.97	0.23	89

other with a low victimization rate, low offence rate, and low resident offender rate.

Summary

We can say that two elements of travel contribute to the patterns outlined in Table 16.1:

1 Offenders tend not to travel far to offend, hence the localized nature of victimization in the areas they live.
2 Victimization of those residents in low offender rate areas often occurs away from the area where they live, as travel brings them into contact with motivated offenders, and very few offenders journey into their residential area. Many victimizations occurred in 'shared' spaces such as the city centre.

Support for the above hypothesis is provided by the different types of offence typically suffered by residents of the two areas; in Area 1 (the high crime area) domestic burglary dominates, whilst residents of Area 2 (the low crime area) are much more likely to be victims of automobile-related crime. One offence is linked to where one lives; the other, in theory, can occur anywhere one drives.

Such findings are not limited to one country or model of transport. Studies of GIS-based analysis of mugging in Chicago and the Bronx, New York City, by Block and Block (1997) indicated that many muggings occur on the street in close proximity to subway stations or bus stops (although not within the more heavily patrolled stations themselves). The logical conclusion is that the victim is making a pedestrian journey from his or her place of employment to mass transit (or to a parking lot) when intercepted by the armed robber who having made a journey via subway or bus to his place of "predation" is then on the look-out for a satisfactory victim. Research by Leipnik and Paulsen (1997) on 10 years of homicides in Houston, Texas, indicated that while relatively few homicides occurred where the victim was white or Hispanic and the killer black, a singular pattern did emerge in the inter-racial cases. The vast majority of such inter-racial homicides occurred in neighborhoods that census data indicated were predominantly black. This indicated a probable "journey to crime" of the victim. In this instance travel by automobile into "alien," predominantly black neighborhoods initiated the conditions for victimization (murder) to occur. In all probability, the journey of the victim was to purchase narcotics or solicit the services of prostitutes. Although instances of "carjackings" where the victim's had the misfortune to stop for gas, directions, or to drop off dry cleaning in a predominantly black area also occurred.

Notes

1 A full account of the findings briefly described here can be found in Wiles and Costello (2000).
2 Work is currently underway examining patterns of movement in Sheffield for the period 1995–2000.
3 Here targets obviously means mobile items such as motor vehicles.
4 Adapted from Tables 3.12 and 3.13, Wiles and Costello, 2000: 22–23.
5 A council estate 2 miles south-east of the city center; it has a poor reputation in the city and is one of its most deprived communities.
6 An exclusive middle-class suburb 2.5 miles west of the city center.

References

Block, R. L. and C. R. Block (1997) Risky places in Chicago and the Bronx: Robbery in the environs of rapid transit stations. Paper presented to the Spatial Analysis of Crime Workshop, Hunter College, New York, NY.

Leipnik, Mark R. and Derek Paulsen (1997) GIS based analysis of demographic factors related to occurrence of homicides over a ten-year period in Houston, Texas. Paper presented at symposium, Exploring the Future of Crime Mapping. A National Symposium on the Use of GIS in Criminal Justice Research and Practice, October, Denver, Colorado.

Pyle, G. (1974) The spatial dynamics of crime. Department of Geography Research Paper, no. 159. Chicago: University of Chicago Press.

Wiles, P. and A. Costello (2000) The road to nowhere: The evidence for travelling criminals. Home Office Research Study, no. 207. London: Home Office.

17 Aerial photography and remote sensing for solving crimes

Joseph Messina and Jamie May

Remotely sensed imagery (including in its broadest sense; aerial photography and aerial videography as well as satellite based systems) is finding increasing application within law enforcement's use of geo-spatial technologies. This is due largely to the increasing availability of digital imagery, particularly 1-meter or better resolution digital ortho-photography and to improvements in the processing power and reductions in the cost of the computers needed to manage this imagery. In general, aerial imagery finds applications in providing a context to the "spaghetti" of street centerlines that are the common base-maps employed in crime mapping and analysis. Thus, an aerial image can show the vegetated areas where residential burglars may lurk, can show that paths, in addition to streets, are available for the movement of criminals, can show that vacant lots are present where bodies or stripped cars may have been dumped. In the context of a specific case, high-resolution aerial imagery can reveal that at the specific location where a murder victim's body was recovered on the bank of a river near Abbotsford, British Columbia there was a walking path which in turn led to a parking area visible from some nearby homes but not from others, due to vegetation. In at least one case in the state of Maine, aerial photography taken for highway engineering, actually showed on subsequent close examination the presence of the body of a murder victim, thus refuting a medical examiner's estimate of time of death and negating the alibi of the person who ultimately confessed to the killing. These specific examples aside, aerial imagery can be a wonderful source of data to help fill in the blank spaces in a GIS and thus make the line work intelligible for laymen, whether they be members of the community or jurors.

(Editor)

Law enforcement activities are increasingly dependent upon technology. One of the most significant developments of the last 30 years is the emergence of satellite and airborne remote sensing systems. While remote sensing can broadly be defined as the acquisition of information about an object without direct contact and can include visual and aural systems, this chapter focuses on the use of visual sensors mounted on airborne or spaceborne platforms. Applying these technologies in a knowledgeable manner first requires a basic understanding of the characteristics of the sensors, data formats, and tradeoffs. This chapter is organized by first providing an example of an application. This is followed by a discussion of the technical aspects of airborne and spaceborne remote sensing systems. The chapter concludes with a discussion of purchase and use considerations.

A case of carjacking

On January 18, 2000 at 1:35 a.m. a 27-year-old man was sitting in his 1997 Jeep Wrangler at a stoplight in Overland Park, Kansas. As he was waiting for the light to turn green, he noticed a 1990s model blue Cadillac approaching him from the rear. Suddenly, the Cadillac bumped him. As the victim looked back, he observed two males inside the vehicle. The victim immediately exited his vehicle to inspect the damage. The passenger of the Cadillac exited the vehicle and approached the victim. The passenger struck the victim in the jaw, entered the victim's Jeep and drove off with the driver of the Cadillac following behind. The victim was able to run to a nearby residence where he called 911.

Information on the carjacking was broadcast immediately. A short time later, a neighboring police department notified the reporting officer that a person matching the suspect description had been stopped and was being detained nearby. The victim was driven to that location and was able to identify the individual as the person responsible for the carjacking. The stolen vehicle was later recovered at another location.

Detective Greg Wilson in the Crimes Against Persons Unit of the Investigation Division approached the Crime Analysis Unit and requested their help in preparing a map to be used for trial. The detective felt that it was necessary to create a map so the jury could easily visualize and therefore better comprehend the series of events that had taken place the night of the incident and the locations at which those events had occurred. Since the creation of the Crime Analysis Unit in 1996, the unit has created numerous GIS-based maps for trial. Most of these maps have been at the request of the prosecution and have received favorable review.

It was decided that for this particular case an aerial photograph would be best to depict the three main locations. They were (1) the incident location, (2) the location where the suspect was apprehended, and (3) the location where the victim's vehicle was recovered. Not only was the aerial photograph going to be used as an aid to visualization, it was also going to be used to help establish a timeline showing that it was possible for the suspect to have traveled from "Point A" to "Point B" in the time from commission of the crime to his apprehension.

ArcView GIS was used to create the map. By using the city's street centerline file, the analyst found the area where the incident occurred. Next, the aerial photo was scanned and imported into ArcView. As mentioned earlier, three locations had to be plotted. Using the "locate address" tool, each specific address was identified and labeled. Several of the city's major streets were also labeled for reference. Before the map could be finalized, the detective verified that each location depicted on the map was the actual location where the events had occurred. After verification, the map was printed out at the size of 34 × 44 inches and was placed on foam board for presentation in court (Figure 17.1).

With the aid of the map, the prosecution was able to help recreate the scene and provide the parties involved (jurors, attorneys, officers, etc.) with a better visualization of the area. The aerial map also eliminated the need for the jurors to be taken to the multiple locations where events had occurred. The prosecutor found the map to be a key tool in proving the sequence of events could have happened in the specified time frame, thus, disproving the defense's contention that the sequence of events could not have occurred in the short time frame presented by the police. The saying that we have all frequently heard, "a picture (map) is worth a thousand words," was once again proven true. As in previous court cases, large-scale crime scene maps have aided in the prosecution of the offender. In this case, the suspect, a repeat violent offender, was sentenced to 19 years in prison for the carjacking.

Technical aspects

The example presented illustrates only a simple application of aerial photographic data. To understand more complex use of this technology, it is important to understand some of the basic terminology. *Image* is used as the general term for all digital visual products. In the previous example, the aerial photograph was scanned and orthorectified (by the provider) to permit easy integration with the GIS software. *Orthorectification* is the process of correcting the imagery for variation in the position of the sensor/platform, for convergence of light, for distortion due to curvature of optical lenses and, more commonly, for irregularity in the terrain. *Platform* is the vehicle (helicopter, airplane, or satellite) carrying the *sensor* (camera – digital or film-based).

Understanding the differences in sensor capabilities is integral to choosing data best suited for a particular application. To assess the characteristics and tradeoffs involved in data acquired from different sensors, it is necessary to have a working knowledge of the characteristics or resolutions by which data are judged. The four types of resolution are spatial, spectral, radiometric, and temporal, and are each discussed below in the context of remotely sensed data applicable to law enforcement applications.

Spatial resolution

Spatial resolution is most easily understood as how large an area on the ground is represented by a single pixel in the image. This concept easily translates into the identification of the smallest features on the ground visible on the image. For example, Landsat Thematic Mapper (TM) data

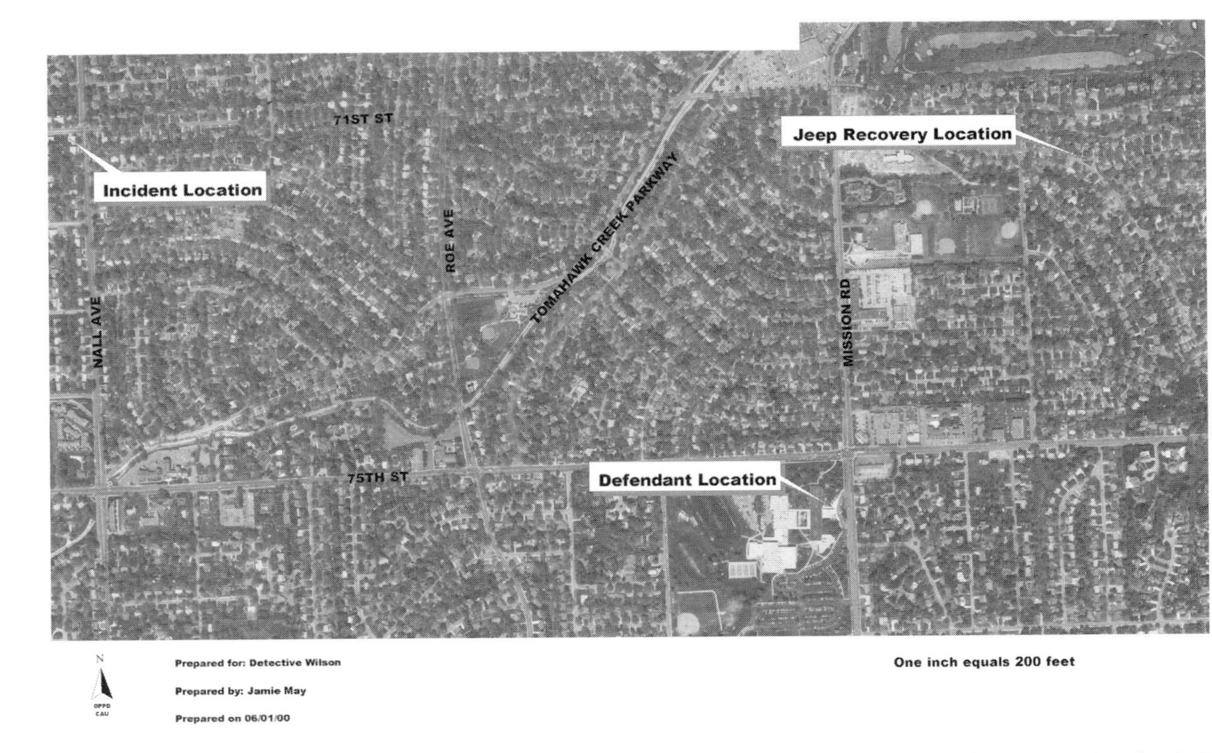

Figure 17.1 The juxtaposition between the incident, apprehension, and recovery locations, Overland Park, Kansas Police Department.

imagery has a spatial resolution of 30 m, meaning that each pixel represents a 30 × 30 m area on the ground. Thus, only relatively large buildings can be identified individually.

When considering spatial or any other type of resolution, it is imperative to consider the purpose of the application; that is, the characteristics of the object one is trying to detect strongly dictate the appropriateness of a given resolution specification. For instance, an analyst looking for large marijuana fields might find 30-m Landsat data to be of sufficient spatial resolution. However, an analyst or law enforcement officer looking for local isolated patches of marijuana embedded within small forest clearings might miss these small patches using 30-m resolution data. Researchers new to remote sensing might immediately assume that one should always use the finest spatial resolution data available for any given project. However, as spatial resolution increases, so too does the size of the data set and therefore the computation power required and costs associated with acquisition and processing. Furthermore, a tradeoff always exists with coverage on the ground and spatial resolution. This is due in large part to data transfer rates.

Landsat Thematic Mapper with its 30-m spatial resolution (15 m with Landsat 7 and ETM or Enhanced Thematic Mapper) is generally the lowest spatial resolution of use in law enforcement activities. However, there are other satellite sensors with higher spatial resolution, including SPOT Multispectral (20 m) and SPOT Panchromatic (10 m), the Indian Remote Sensor (IRS) with 8-m spatial resolution and the new Ikonos with 1–4 m spatial resolution.

All the aforementioned systems are spaceborne. By moving into the atmosphere, the options dramatically increase, but often with an attendant increase in cost and reduced options for collections as well as legal considerations. The airborne systems available today can image features as small as a centimeter if at a low altitude. However, the collection and processing demands can be significant. Active (non-optical) sensors, like RADAR, also exhibit an array of spatial resolutions that vary depending upon mode. Because of the tradeoff between scene size and spatial resolution, law enforcement officials should possess a working knowledge of which resolutions and media are best suited for a particular objective.

Spectral resolution

The second type of resolution of concern in data selection is spectral resolution, which concerns the number and width of the bands of the electromagnetic spectrum imaged by the sensor. Optical sensors record in the visible and infrared portions of the electromagnetic spectrum. Within that range, the sensor may break up the reflected energy it senses into any number of bands. For example, a sensor might have detection capabilities for three bands, one that records energy reflected in the visible green, another in the visible red and a third that records in the infrared spectrum. This example is a common spectral resolution/band combination (e.g. SPOT, and in a single image color infrared aerial photography). Newer developments in remote sensing technology have led to hyperspectral sensors with very high spectral resolution, where, for example, the sensors may detect upwards of 200 bands covering the same portion of the spectrum as traditional sensors, making each band very narrow. The concern with bandwidth has to do with the need to discern features in the landscape. With very few exceptions, every object reflects some amount of energy in various portions of the spectrum. The typical reflectance of an object can be quantified through a set of spectral signatures that can then be used as a type of key to recognize similar features elsewhere. The first application of multispectral imagery was in the Second World War, where color infrared photography was used to identify tanks etc. hidden under camouflage netting.

For example, suppose that an analyst were interested in locating a large patch of marijuana. Building the scenario further, suppose that we know that the patch is greater than 10 acres and totally enclosed within a cornfield. With this information, we know that at least two vegetation signatures need to be distinguished. From established spectral signatures (via previous research), the analyst knows that both types reflect at low levels in the blue (0.4–0.5 μm) portion of the electromagnetic spectrum and at medium levels in the lower portion of the near infrared (NIR) portion (0.7–0.8 μm). However, dry corn has a much higher reflectance in the red portion (0.6–0.7 μm) and a much lower reflectance in the upper portion of the NIR (0.8–1.1). Consequently, the ability to differentiate among visible (blue, green, and red) bands would be important, although the inclusion of the blue band would not be vital, since dry corn and green marijuana are not readily spectrally separable (able to be distinguished) in that band. It would also be important to be able to distinguish among different portions of the NIR section of the electromagnetic spectrum. Thus, the analyst should look for data from a sensor that (1) has at least two NIR bands (or that the NIR band covers only the upper portion of NIR); (2) separates the red band from other visible portions of the spectrum, and; (3) that has the spatial resolution necessary to capture the 10 acre plot of marijuana. However, suppose the analyst needs to distinguish among features in an urban environment (e.g. to detect illegal dumping), then higher spectral resolution data (perhaps with the blue band or multiple, narrow visible bands) might become necessary. Ultimately, to most efficiently select the appropriate satellite dataset, it is important to be familiar with the target of interest as well as the physical site and situation of the region. To aid in this selection, law enforcement officers and consultants should familiarize themselves with the basic applications of various wavelengths: (1) visible blue for differentiating soil from vegetation or locating objects (such as dumped automobiles) in shallow water; (2) visible green for discerning health of vegetation; (3) visible red for discriminating among vegetation species and distinguishing water from land; (4) near infrared for general vegetation mapping, vegetation species discrimination, and vegetation health/phenological cycle (e.g. growth, maturation, senescence); (5) mid-infrared for locating water/land boundaries and detecting moisture in vegetation or soil. Further, it should be remembered that most often bands are used in combination, whether stacked for viewing purposes or used in combination to calculate indices.

Radiometric resolution

Radiometric resolution is the third type of resolution to consider in data selection, and refers to the sensitivity to brightness level. That is, how many brightness levels (in any given spectral band) can the sensor detect. Consider an example using a pencil and paper. If you draw a picture on the paper you have created "1-bit" line; it has two shades only, therefore it can be considered to have "1-bit" radiometric resolution. If you were to shade in one portion of the picture then you've created a level of shading and radiometric resolution. Refer to the marijuana and corn cited previously. Dry corn and marijuana both reflect energy in the red portion of the electromagnetic spectrum; simply having the red portion separated from the blue and green portions of the spectrum would not help to identify whether an area was predominantly dry corn or green marijuana. However, based on spectral reflectance characteristics, it is known that dry corn has a much higher (brighter) reflectance in the red band than marijuana. The ability to distinguish enough levels of brightness to detect this difference is radiometric resolution, and works in conjunction with spectral resolution. With satellite imagery, this resolution is normally referred to in bits; 1-bit for the line art example. For Landsat TM there is 8-bit resolution, implying that each band distinguishes among 256 (2^8) levels of brightness. The radiometric resolution of scanned aerial photography is dependent upon the capabilities of the scanner and to some extent the quality of the photography.

Temporal resolution

The fourth and final type of resolution is temporal, and refers to the time interval between periods of observation in the same area for a given sensor. Also referred to as a repeat cycle, this type of resolution is especially important when performing change detection analysis or when intra-annual seasonality is important for the phenomenon of interest. For instance, monitoring suspected areas of marijuana production during the growing season would require data for the same area at multiple times per year. The Ikonos and SPOT HRV sensors are examples of sensors designed to take advantage of temporal resolution. These sensors are pointable, meaning that they can capture imagery from areas not directly underneath them. Faster repeat cycles or higher temporal resolution allow for greater operational flexibility than is normally allowed by a fixed orbital platform. The greatest advantage of airborne platforms like airplanes and helicopters is that they can be scheduled to collect data any time needed, given limitations of cloud cover.

Trade offs

Evaluating the four types of resolution for any particular project is a significant challenge. No single sensor maximizes every type of resolution because of the inherent tradeoffs involved in the different types of sensors. Historically, the primary limitation in all remote sensing sensor designs was the ability of the platform to send the collected data to the earth, a communications issue. For example, increasing spatial resolution necessitates a smaller ground coverage footprint. Landsat TM at 30 m covers an area roughly 9 times that of SPOT 10-m Panchromatic data. This footprint decrease also impacts the temporal resolution. The Landsat repeat cycle of 16 days is significantly shorter than the 26-day cycle of SPOT. The higher spatial resolution sensors ameliorate that problem by offering pointable sensors. Similarly, increasing radiometric resolution also requires a larger volume data stream and leads to a decrease in one of the other types of resolution. IRS-1D panchromatic data has a spatial resolution of 8 m but only 6-bit radiometric resolution. Generally, increasing any one type of resolution increases file sizes, data processing requirements, and costs. Therefore minimum resolution requirements should be set and prioritized as early in a project as is possible.

Purchase and use considerations

Aerial photography is commonly available from the Federal government through the United States Geological Survey, the National Aerial Photography Program Office in Salt Lake City and the EROS Data Center in Sioux Falls South Dakota. The primary data product of interest is the Digital Orthophoto Quad; a scanned and orthorectified aerial photograph. Complete coverage of the United States exists and these data can be easily imported into most GIS programs.

A checklist can be used to help an analyst select appropriate data or manage contacts with a data provider.

Identifying appropriate data: target of interest

1 Urban or Rural, vegetation or soils/rock: this affects the type of imagery desired including spectral resolution choices.
2 Size of area: the larger the area of interest the more data will be needed. This is a cost consideration as well as processing and availability consideration.
3 Time of year: this is separate from temporal resolution. Time of year consideration is important because it affects the image potential due to sun angle as well as influencing vegetation phenology. In the winter, a better image choice might be panchromatic or simple black and white imagery, as spectral differences will be minimized also trees may be in a "leaf-off" condition.

4 How quickly and how often is the imagery required? Many sensors collect only once a month or more often only by scheduling. It is unlikely that satellite imagery will be available for a specific date.

Information for data providers

1 Location: by latitude and longitude, UTM coordinates, state plane coordinates, etc. and within the United States, place name and zip code are acceptable.
2 Time that the imagery was acquired: be as specific as possible.
3 Product: most data providers carry multiple data lines. It is important to be prepared to be specific regarding the choice of a product.
4 Processing level: the data must be made suitable for use in the software package, unless you have significant computing power and image processing software such as ERDAS Imagine, or ARCView Image Analyst.

While there are many sources of information regarding the basic use of camera systems, aircraft, and helicopters, much of this work has focused on real-time traffic and perpetrator tracking. Remote sensing from spaceborne systems serves other purposes and can fulfill other mission requirements. From detection of drug production activities to determination of illegal dumping locations, the applications of the data are more a function of training and acceptance within the law enforcement community than with limitations in the data or systems. As GIS and image processing software become easier to use and remotely sensed imagery less expensive, it is inevitable that more law enforcement agencies will begin to use satellite imagery and aerial photography for law enforcement activities.

Part III

Appendices

A Towards a lexicon of criminology and geography

25 useful terms

Donald P. Albert and Mark R. Leipnik

The interaction between law enforcement and geography has made tremendous contributions to the lexicons of these vibrant disciplines. The fact is that geography must be taken into account when analyzing most human phenomena, including patterns of criminal offenses. This linkage across disciplines requires the development of a unique lexicon through the adaptation of old terms and the coinage of new ones. Some of these terms evolved as criminologists began using computer mapping and geographic information systems to manage and analyze spatial attributes associated with crime data. The following is a collection of 25 useful terms that have found their place in the joint lexicon of criminologists and geographers. This glossary is not intended to be complete or comprehensive, but it does reflect the recent cross-disciplinary connections that have brought criminologists and geographers together.

Glossary

Activity space The area within which offenders move during the regular round of their activities.

Anchor points Focus points such as residences, workplaces, criminal markets, and other hangouts that are important to criminals.

Appearance disorder The average number of disorders (broken windows, graffiti) per building in a geographic unit (blocks).

Circle hypothesis The hypothesis that an offender's home base lies within a circle drawn around a line (diameter) that has as its endpoints the offender's two most distant crime locations.

Commuter offenders Criminals who live in one area but "work" in another.

Crime-gradient The construction of contour maps where isopleths, or lines, connect points of equal crime rates.

Crime wave The spatial and temporal clustering of crime incidents.

Exporter A criminal who generates a substantial number of offenses outside of his own immediate neighborhood.

Geoforensics The use of geographic concepts and themes during the analysis of crime patterns.

Geographic profiling The geographic analysis of a serial offender's crime locations for the purpose of zeroing in on the perpetrator's neighborhood.

"Hit" rate The percentage of geographic locations (e.g. addresses) correctly matched with geographic coordinates.

Hot spot area The place of highest crime density, as revealed through the use of statistical cluster analysis and often highlighted by circles, ellipses, or other geometric shapes.

Journey-to-crime The distance from an offender's home to the offense location.

Location Quotients in Crime (LQC) The ratio of the rate of crime in a sub-unit to the rate of crime of the complete geographic unit.

Manhattan distance The street distance between point a and b that requires going around a corner to reach point b.

Mapless mapping Mapping criminal data, using latitude and longitude or some other coordinate system, without reference to geographic boundaries.

Marauder A rapist who commits offenses in a more or less uniform pattern within a circle encompassing his base of operation. The circle is "domocentric" if that base of operation is the rapist's residence.

Place The amount of the built environment that can be viewed from a given surface position with the unaided eye.

Safety zone The distance between a rapist's residence and the location of his closest offense.

Sphere of concern This method uses vectors drawn from the place of abduction *to* the place of recovered bodies. Individual vectors are aggregated to define a radius that is used to construct a circle or sphere of concern.

Stationary fallacy Suggests that when data on offenses are combined over different time periods (i.e. combining daytime and nighttime offenses), possibilities exist for the identification of false clusters. For example, if thefts are a daytime phenomenon and assaults a nighttime phenomenon, aggregating these might produce a cluster representing unrelated crimes.

Target backcloth The spatial and temporal distribution of "suitable" crime targets or victims. (The quotations around suitable indicates that this term reflects the perspective of the offender.)

Territorial, geographic, or spatial displacement The relocation of criminal activities in response to decreasing crime opportunities or increasing chance of apprehension. Area restrictions included within bail and probation orders might also stimulate territorial displacement.

Windshield wiper or Pie shape Describes the locations of criminal offenses that are characterized by a distinct directional pattern.

Wheel distance This refers to network distance or the actual distance traveled from points a to b. Wheel distance contrasts with straight-line or Euclidean distance.

B Crime mapping resources

Mark R. Leipnik and Donald P. Albert

Books

Block, C., M. Dabdoub, and S. Fregly (1995) *Crime Analysis through Computer Mapping*. Washington, DC: Police Executive Forum.

Goldsmith, V., P. G. McGuire, J. H. Mollenkopf, and T. A. Ross (2000) *Analyzing Crime Patterns: Frontiers of Practice*. Thousand Oaks, CA: Sage Publications.

Harries, K. (1999) *Mapping Crime: Principle and Practice* Washington, DC: National Institute of Justice.

Hirschfield, Alex and Kate Bowers (eds.) (2001) *Mapping and Analysis of Crime Data*. London: Taylor and Francis.

La Vinge, N. and J. Wartell (1998) *Crime Mapping Case Studies: Successes in the Field*. Washington, DC: Crime Mapping Research Center and Police Executive Research Forum.

Weisburd, D. and T. McEwen (1997) *Crime Mapping and Crime Prevention*. Monsey, NY: Criminal Justice Press.

Bibliography

GIS/GPS in Law Enforcement Master Bibliography, by Donald P. Albert and Susan D. Strickland, is available for free download from the Police Executive Research Forum's web site at http://www.policeforum.org.

Government and educational organizations

Crime Mapping Research Center, US Department of Justice. National Institute of Justice. Washington, DC
http://www.ojp.usdoj.gov/cmrc

Geospatial Information Technology Association. Washington, DC
http://www.gita.org

National Center for Geographic Information and Analysis, University of California at Santa Barbara. Santa Barbara, California.
http://www.ncgia.ucsb.edu/

Police Executive Research Forum. Washington, DC
http://www.policeforum.org
The Police Foundation. Washington, DC
http://www.policefoundation.org
Urban and Regional Information Systems Association. Washington, DC
http://www.urisa.org

Data sources

Crime Mapping Research Center
http://www.ojp.usdoj.gov/cmrc
Geographic Data Technology, Inc.
http://www.geographic.com/
Ordinance Survey of the United Kingdom
http://www.ordsvy.uk/
Teleatlas, Inc.
http://teleatlas.com/
The Geography Network
http://www.geographynetwork.com
U.S. Census Bureau
http://www.census.gov/

Software vendors/consultants

Autodesk/Mapguide
http://autodesk/mapguide
Crime Prevention Analysis Lab, Inc (CPAL)
http://WWW.crimepatterns.com
Environmental Criminology Research, Inc (ECRI)
http://www.ECRIcanada.com
The Omega Group, Inc (CrimeView)
http://www.theomegagroup.com/
4th Watch Systems, Inc.
www.4th watch.net
National Law Enforcement and Corrections Technology – Rocky Mountain Region. Denver, Colorado. http://www.nlectc.org/nlectcrm/
Illinois Criminal Justice Information Authority. Chicago, Illinois. http://www.icjia.state.il.us/public/index.cfm
Environmental Systems Research Institute, Inc. (ArcView, ARC/INFO) http://www.esri.com/
GE-Smallworld. http://www.smallworld.us.com/
Intergraph Corporation (Geomedia) http://www.intergraph.com/
MapInfo Corporation http://www.mapinfo.com/

C Master bibliography

Compiled by Donald P. Albert

A

Agung, A. (1997) Crime hot spot analysis and dynamic pin map. *Proceedings, 1997 Environmental Systems Research Institute International User Conference*. Available at http://www.esri.com/library/userconf/archive.html.

Alexander, M., E. Groff, L. Hibdon (1997) An automated system for the identification and prioritization of rape suspects. *Proceedings, 1997 Environmental Systems Research Institute International User Conference*. Available at http://www.esri.com/library/userconf/archive.html.

Alexander, R. (1999) With GIS mapping software, local police gaining new views. *The Boston Globe*, 3rd edn, (November 21): 5.

Allen, G. (1993) Advances in automatic vehicle location technology. *Police and Security News* 9(5): 3, 47–50.

Alvaredo, F. and M. Gomez (1992) GIS supports summer Olympic games security. *GIS World* 5(6): 58–61.

Anderson, D. (1990) Seattle and Tacoma PDs automated crime analysis. *National F. O. P. Journal* 19(2): 50–2.

Anderson, D. (1990) Seattle, Tacoma automated crime analysis. *American City and County* 105 (July): 52.

Anderson, M. G. (1996) GPS used to track criminals. *GIS World* 9(8): 15.

Anonymous (1991) Exploring new worlds with GIS. *IBM Directions* 5(2): 12–19.

Anonymous (1992) Illinois uses computers to identify criminal hot spots. *State Legislatures* 18(12): 7.

Anonymous (1998) The IACP LEIM Section: Finding order in the complex world of new technology. *Police Chief* 65(3): 14.

B

Baker, T. E. (1999) Supergangs – or organized crime? *Law and Order* 47(10): 192–7.

Baker, W. T. (1997) ALERT: Police vehicle technology for the 21st century. *Police Chief* (September): 23–33.

Barnes, G. C. (1995) Defining and optimizing displacement. In: J. E. Eck and D. Weisburd (eds), *Crime and Place*. Monsey, NY: Criminal Justice Press; and Washington, DC: Police Executive Research Forum, pp. 95–113.

Barr, R. and K. Pease (1990) Crime placement, displacement and deflection. In: M. Tonry and N. Morris (eds), *Crime and Justice: A Review of Research*, vol. 12. Chicago: University of Chicago Press.

Bellucci, C. (1995) DMAP in Jersey City: Implementing a technological revolution. In: C. R. Block, M. Dabdoub, and S. Fregly (eds), *Crime Analysis Through Computer Mapping*. Washington, DC: Police Executive Research Forum, pp. 195–9.

Bennett, T. (1995) Identifying, explaining, and targeting burglary hot spots. *European Journal on Criminal Policy and Research* 3(3): 113–23.

Bennett, W. D., A. Merlo, and K. K. Leiker (1987) Geographical patterns of incendiary and accidental fires. *Journal of Quantitative Criminology* 3(1): 47–64.

Block, C. R. (1995) STAC hot-spot areas: A statistical tool for law enforcement decisions. In: C. R. Block, M. Dabdoub, and S. Fregly (eds), *Crime Analysis Through Computer Mapping*. Washington, DC: Police Executive Research Forum, pp. 15–32.

Block, C. R. (1997) The GeoArchive: An information foundation for community policing. In: D. Weisburd and J. T. McEwen (eds), *Crime Mapping and Crime Prevention*. Monsey, NY: Criminal Justice Press, pp. 27–81.

Block, C. R. and M. Dabdoub (1993) *Workshop on Crime Analysis Through Computer Mapping Proceedings*. Chicago: Illinois Criminal Justice Information Authority.

Block, C. R. and L. A. Green (1994) *The Geoarchive Handbook: A Guide for Developing a Geographic Database as an Information Foundation for Community Policing*. Chicago: Illinois Criminal Justice Information Authority.

Block, C. R., M. Dabdoub, and S. Fregly (eds) (1995) *Crime Analysis Through Computer Mapping*. Washington, DC: Police Executive Research Forum.

Block, R. L. (1995) Geocoding of crime incidents using the 1990 TIGER File: The Chicago example. In: C. R. Block, M. Dabdoub, and S. Fregly (eds), *Crime Analysis Through Computer Mapping*. Washington, DC: Police Executive Research Forum, pp. 189–93.

Block, R. L. (1995) Spatial analysis in the evaluation of the "CAPS" community policing program in Chicago. In: C. R. Block, M. Dabdoub, and S. Fregly (eds), *Crime Analysis Through Computer Mapping*. Washington, DC: Police Executive Research Forum, pp. 251–8.

Block, R. L. and C. R. Block (1995) Space, place, and crime: Hot spot areas and hot places of liquor-related crime. In: J. E. Eck and D. Weisburd (eds), *Crime and Place*. Monsey, NY: Criminal Justice Press; and Washington, DC: Police Executive Research Forum, pp. 145–84.

Block, R. L. and C. R. Block (1997) Risky places in Chicago and the Bronx: Robbery in the environs of rapid transit stations. Paper presented to the Spatial Analysis of Crime Workshop, Hunter College, New York, NY.

Bowman, B. A. (1989) Commemorating the 30th anniversary of airborne law enforcement in LA (Los Angeles). *Police Chief* 56(2): 17–18.

Bowman-Jamieson, D., P. Drummy, and P. Scanlon (1999) A regional approach to crime mapping and the Web. *Proceedings, 1999 Environmental Systems Research Institute International User Conference*. Available at http://www.esri.com/library/userconf/archive.html.

Bowron, S. F. M. and R. J. Ruprecht (1981) Fixed wing aircraft in support of police operations – An introductory report of the work of the Hampshire constabulary air support unit. *Police Research Bulletin* 37: 38–45.

Bradshaw, T. L. (1999) Gunfire location system helps stop crime in urban areas. *Proceedings, 1999 Environmental Systems Research Institute International User Conference*. Available at http://www.esri.com/library/userconf/archive.html.

Brantingham, P. J. and P. L. Brantingham (1981) *Environmental Criminology*. Beverly Hills, CA: Sage Publications.

Brantingham, P. J. and P. L. Brantingham (1984) *Patterns in Crime*. New York, NY: Macmillan.

Brantingham, P. L. and P. J. Brantingham (1995) Location quotients and crime hot spots in the city. In: C. R. Block, M. Dabdoub, and S. Fregly (eds), *Crime Analysis Through Computer Mapping*. Washington, DC: Police Executive Research Forum, pp. 129–49.

Brantingham, P. L. and P. J. Brantingham (1997) Mapping crime for analytic purposes: Location quotients, counts and rates. In: D. Weisburd and T. McEwen (eds), *Crime Mapping and Crime Prevention*. Monsey, NY: Criminal Justice Press, pp. 263–88.

Brassel, K. E. and J. J. Utano (1979) Linking crime and census information within a crime mapping system. *Review of Public Data Use* 7(3/4): 15–24.

Brassel, K. E. and J. J. Utano (1979) Mapping from an automated display system. *Professional Geographer* 31: 191–200.

Brown, D. E. (1992) Drugs on the border: The role of the military. *Parameters* 21(1991/1992): 50–9.

Brown, D. E. and H. Liu (1999) A new approach to spatial-temporal criminal event prediction. *Proceedings, 1999 Environmental Systems Research Institute International User Conference*. Available at http://www.esri.com/library/userconf/archive.html.

Brown, G. W. (1994) *What Impact will Personal Position Location Technology have upon the Management and Administration of Mid-Sized Law Enforcement Organization by the Year 2000*? Sacramento: California Commission on Peace Office Standards and Training.

Brown, M. A. (1982) Modeling the spatial distribution of suburban crime. *Economic Geography* 58(3): 247–61.

Brown, S., D. Lawless, X. Lu, and D. J. Rogers (1998) Interdicting a burglary pattern: GIS and crime analysis in the Aurora Police Department. In: N. La Vigne and J. Wartell (eds), *Crime Mapping Case Studies: Successes in the Field*. Washington, DC: Police Executive Research Forum, pp. 99–108.

Buerger, M. E., E. G. Cohn, and A. J. Petrosino (1995) Defining the "hot spots" of crime: Operationalizing theoretical concepts for field research. In: J. E. Eck and D. Weisburd (eds), *Crime and Place*. Monsey, NY: Criminal Justice Press; and Washington, DC: Police Executive Research Forum, pp. 237–58.

Bulkeley, D. (1993) High-tech tools to help crack crimes. *Design News* 48(49): 27–8.

Burka, J. C., A. Mudd, D. Nulph, and R. Wilson (1999) Breaking down jurisdictional barriers: A technical approach to regional crime analysis. *Proceedings, 1999 Environmental Systems Research Institute International User Conference*. Available at http://www.esri.com/library/userconf/archive.html.

Buslik, M. and M. D. Maltz (1997) Power to the people: Mapping information sharing in the Chicago Police Department. In: D. Weisburd and T. McEwen (eds), *Crime Mapping and Crime Prevention*. Monsey, NY: Criminal Justice Press, pp. 113–30.

C

Call, R., R. Mayer, and R. Baird (1989) From punch cards to computers: An evolution in crime analysis. *Police Chief* 56(6): 37.

Cameron, B. W. (1999) Technology improves crime statistics. *Law and Order* 47(3): 30–2.

Campbell, G. (1992) GIS in the police environment. In: J. Cadoux-Hudson and I. Heywood (eds) *Geographical Information 1992/3: The Yearbook of the AGI*. London: Taylor & Francis, pp. 114–8.

Canter, P. R. (1995) State of the statistical art: Point pattern analysis. In: C. R. Block, M. Dabdoub, and S. Fregly (eds), *Crime Analysis Through Computer Mapping*. Washington, DC: Police Executive Research Forum, pp. 151–60.

Canter, P. R. (1997) Geographic information systems and crime analysis in Baltimore County, Maryland. In: D. Weisburd and J. T. McEwen (eds), *Crime Mapping and Crime Prevention*. Monsey, NY: Criminal Justice Press, pp. 157–90.

Canter, P. R. (1998) Baltimore County's autodialer system. In: N. La Vigne and J. Wartell (eds), *Crime Mapping Case Studies: Successes in the Field*. Washington, DC: Police Executive Research Forum, pp. 81–92.

Carnaghi, J. and J. T. McEwen (1970) Automatic pinning. In: S. I. Cohn and W. E. McMahon (eds), *Law Enforcement, Science, and Technology*, vol. III. Chicago: Illinois Institute of Technology Research.

Cheetham, R. and K. Switala (1998) Police at play: What if redistricting plans were as addictive as Solitarire™? *Proceedings, 1998 Environmental Systems Research Institute International User Conference*. Available at http://www.esri.com/library/userconf/archive.html.

Chicca, E. L. (2000) Developing automated systems to track false alarms for local jurisdictions. *Proceedings, 1997 Environmental Systems Research Institute International User Conference*. Available at http://www.esri.com/library/userconf/archive.html.

Clarke, R. V. (ed.) (1992) *Situational Crime Prevention: Successful Case Studies*. New York, NY: Harrow and Heston.

Claypool, D. W., S. P. Zietlow, B. K. Gilbert, K. A. Creager. J. L. Tri, Y. Novick, and J. Silver (1995) *Proceedings, Counter Law Enforcement: Applied Technology for Improved Operational Effectiveness International Technology Symposium*. Washington, DC: US Executive Office of the President, Part 2, 17-1–17-9.

Clontz, K. A. (1997) Spatial analysis of residential burglaries in Tallahassee, Florida. *Proceedings, 1997 Environmental Systems Research Institute International User Conference*. Available at http://www.esri.com/library/userconf/archive.html.

Clontz, K. A. (1998) Using Atlas Pro analysis for examining the relationship between commercial land use and burglary. *Proceedings, 1998 Environmental Systems Research Institute International User Conference*. Available at http://www.esri.com/library/userconf/archive.html.

Clontz, K. A. and J. G. Mericle (1999) Community policing: Put your mapping where your mouth is. *Proceedings, 1999 Environmental Systems Research Institute International User Conference*. Available at http://www.esri.com/library/userconf/archive.html.

Conley, J. B. (2000) Community safety information system (CSIS) in Winston-Salem, North Carolina. *Proceedings, 2000 Environmental Systems Research Institute International User Conference*. Available at http://www.esri.com/library/userconf/archive.html.

Conley, J. B., E. Groff, and T. Lesser (1999) Using GIS to support community safety: Strategic approaches to community safety (SACS) initiative in Winston-Salem, NC. *Proceedings, 1999 Environmental Systems Research Institute International User Conference*. Available at http://www.esri.com/library/userconf/archive.html.

Cotton, F. B. (1994) *Justice applications of computer animation*. Sacramento: Search Group, Inc.

Curry, G. D. and S. H. Decker (1997) Understanding and responding to gangs in an emerging gang problem context. *Valparaiso University Law Review* 31(2): 523–33.

D

Dadigan, L. and F. Ferguson (1978) Computer mapping aids pinpoint crime and accident trends. *Police Chief* 45(9): 30–32.

DeAngelis, T. B. (2000) GIS answering the why of where. *Police Chief* 67(2): 12.

Devery, C. (1992) Mapping crime in local government area: Assault and break and enter in Waverly. Sydney: New South Wales Bureau of Crime Statistics and Research.

Dighton, D. (1996) Violence of street gangs. *Compiler* 16(2): 4–6.

Duenas, M. (1995) Ada County rewrites the book with GIS. *American City & County* 110(2): 43.

E

Echelberry, N. (1989) Move over, Crockett and Tubbs: Miami has a new crime-fighting weapon. *Police Chief* 56 (Jan.): 20–21, 24–25.

Eck, J. E. (1995) The usefulness of maps for area and place research: An example from a study of retail drug dealing. In: C. R. Block, M. Dabdoub, and S. Fregly (eds), *Crime Analysis Through Computer Mapping*. Washington, DC: Police Executive Research Forum, pp. 277–84.

Eck, J. E. (1997) What do those dots mean? Mapping theories with data. In: D. Weisburd and J. T. McEwen (eds), *Crime Mapping and Crime Prevention*. Monsey, NY: Criminal Justice Press, pp. 379–406.

Eck, J. E. and D. Weisburd (1995) *Crime and Place*. Monsey, NY: Criminal Justice Press; and Washington, DC: Police Executive Research Forum.

Eck, J. E. and D. Weisburd (1995) Crime places in crime theory. In: J. E. Eck and D. Weisburd (eds), *Crime and Place*. Monsey, NY: Criminal Justice Press; and Washington, DC: Police Executive Research Forum, pp. 1–33.

Evans, D. J. and D. T. Herbert (1989) *The Geography of Crime*. London, UK: Routledge.

Evans, D. J. and M. Fletcher (2000) Fear of crime: Testing alternative hypotheses. *Applied Geography* 20 (Dec.): 395–411.

F

Figlio, R. M., S. Hakim, and G. F. Rengert (1986) *Metropolitan Crime Patterns*. Monsey, NY: Criminal Justice Press.

Fortner, R. E. (1998) Computer technology: Mapping the future. *Police* 22(7): 16–21.

Fox-Clinch, J. (1997) Crime and the digital dragnet. *Mapping Awareness* 11(2) March: 22–3.

France, D. L., T. J. Griffin, J. G. Swanburg, J. W. Linemann, G. C. Davenport, V. Trammell, C. T. Armbrust, B. Konratieff, A. Nelson, K. Castellano, and D. Hopkins (1992) Multidisciplinary approach to the detection of clandestine graves. *Journal of Forensic Sciences* 37(6): 1445–58.

Fryer, G. E. and T. J. Miyoshi (1995) Cluster analysis of detected and substantiated child maltreatment incidents in rural Colorado. *Child Abuse and Neglect* 19(3): 363–9.

Fyfe, N. (1991) The police, space, and society: The geography of policing. *Progress in Human Geography* 5(3): 249–67.

G

Geake, E. (1993) How PC's predict where crime will strike. *New Scientist* 140 (Sept.): 17.

Geggie, P. F. (1998) Mapping and serial crime prediction. In: N. LaVigne and J. Wartell (eds), *Crime Mapping Case Studies: Successes in the Field*. Washington, DC: Police Executive Research Forum, pp. 109–16.

Georges, D. (1978) *The Geography of Crime and Violence: A Spatial and Ecological Perspective*. Washington, DC: Association of American Geographers.

Georges-Abeyie, D. E. and K. Harries (eds) (1980) *Crime: A Spatial Perspective*. New York, NY: Columbia University Press.

Glaser, M. B. (1991) GIS for public safety. *Nine-one-one Magazine* 4(3): 26–8.

Goldsmith, V., P. G. McGuire, and J. H. Mollenkopf (2000) *Analysing Crime Patterns: Frontiers of Practice*. Thousand Oaks: Sage Publications.

Gravesen, G. W. (1999) Using laser mapping equipment and procedures. *Law and Order* 47(11): 45–7.

Gray, K. (2000) GIS saving lives. *Eom* 9(3): 14.

Green, L. A. and R. B. Whitaker (1995) *The Early Warning System GeoArchive Codebook: Area Four Project*. Chicago: Illinois Criminal Justice Information Authority.

Grescoe, T. (1996) The geography of crime. *Geographical Magazine* (9): 26–7.

Gresty, B. and K. Taylor (1995) Operation Newburg: A police crackdown on domestic burglaries in Newton-le-Willows, Merseyside. *Focus on Police Research and Development* 6: 18–21.

Groff, E., N. La Vinge, C. Nahbedian, E. Jefferis, M. O'Connell, J. Szaka, and J. Wartell (1998) A multi-method exploration of crime hot spots: An evaluation of the "Repeat Places" Mapping Technique. Paper presented at the Annual Meeting of the Academy of Criminal Justice Sciences. Albuquerque, NM, March 11, 1998.

Grogger, J. and M. Weatherford (1995) Crime, policing and the perception of neighbourhood safety. *Political Geography* 14(6–7): 521–41.

Gwinn, M. (1993) Info bites – computerized public information offers a world of information for fun and profit, but citizen access is fraught with challenges. *The Seattle Times* (February 22): Section B, p. 1.

H

Hakim, S. and G. F. Rengert (1981) *Crime Spillover*. Beverly Hills, CA: Sage Publications.

Harries, K. (1971) The geography of American crime. *Journal of Geography* 70: 204–13.

Harries, K. (1973) Social indicators and metropolitan variations in crime. *Proceedings, Association of American Geographers* 5: 97–101.
Harries, K. (1974) *The Geography of Crime and Justice.* New York, NY: McGraw-Hill.
Harries, K. (1976) Cities and crime: A geographic model. *Criminology* 14: 369–86.
Harries, K. (1978) *Local Crime Rates: An Empirical Approach for Law Enforcement Agencies, Crime Analysts, and Criminal Justice Planners.* Final Report. Grant no. 78-NIJ-AX-0064. Washington, DC: US Department of Justice, Law Enforcement Assistance Administration.
Harries, K. (1980) *Crime and the Environment.* Springfield, IL: Charles C. Thomas.
Harries, K. (1988) Spatial and temporal dimensions of assaults against children in Dallas, Texas, 1980–1981. *Journal of Family Violence* 3(4): 327–38.
Harries, K. (1989) Homicide and assault: A comparative analysis of attributes in Dallas neighborhoods, 1981–1985. *Professional Geographer* 41(Feb.): 29–38.
Harries, K. (1990) *Geographic Factors in Policing*, Washington, DC: Police Executive Research Forum.
Harries, K. (1990) *Serious Violence: Patterns of Homicide and Assault in the U.S.* Springfield, Illinois: Charles C. Thomas.
Harries, K. (1995) A database for the ecological analysis of social stress and violence in small areas. In: C. R. Block, M. Dabdoub, and S. Fregly (eds), *Crime Analysis Through Computer Mapping.* Washington, DC: Police Executive Research Forum, pp. 167–78.
Harries, K. (1995) The ecology of homicide and assault: Baltimore city and county, 1989–91. *Studies in Crime and Crime Prevention* 4: 44–60.
Harries, K. (1997) *Serious Violence: Patterns of Homicide and Assault in America*, 2nd edn. Springfield, IL: Charles C. Thomas.
Harries, K. and R. P. Lura (1974) The geography of justice: Sentencing variations in U.S. judicial districts. *Judicature* 57: 392–401.
Harries, K. and A. Powell (1994) Juvenile gun crime and social stress, Baltimore, 1980–1990. *Urban Geography* 15: 45–63.
Harries, K. and D. Cheatwood (1997) *The Geography of Execution: The Capital Punishment Quagmire in America.* Lanham, MD: Rowman and Littlefield.
Harries, K. and E. Kovandzic (1999) Persistence, intensity, and areal extent of violence against women: Baltimore City, 1992–95. *Violence Against Women* 5: 813–28.
Harris, D. (1985) Sharing crime analysis techniques and information. *Police Chief* 52 (Sept.): 42–3.
Harris, R., C. Huenke, and J. P. O'Connell (1998) Using mapping to increase released offenders' access to services. In: N. La Vigne and J. Wartell (eds), *Crime Mapping Case Studies: Successes in the Field.* Washington, DC: Police Executive Research Forum, pp. 61–6.
Harshaw, L. (1995) Taking the guess out of GIS. *Presentations* 9(7): 14–16.
Hedstrom, P. (1994) Contagious collectives: On the spatial diffusion of Swedish trade unions, 1890–1940. *American Journal of Sociology* 99(5): 1157–79.
Hejazi, F. (1996) Automated redistricting system for law enforcement. *Proceedings, 1996 Environmental Systems Research Institute International User Conference.* Available at http://www.esri.com/library/userconf/archive.html.
Herbert, D. (1982) *The Geography of Urban Crime.* New York: Longman.

Higgins, R. (1999) Computerized mapping gives towns new tool; Using multiple layers of data, GIS has variety of applications. *The Boston Globe*, City Edition-South Weekly (September 12): 1.

Higgins, R. (1999) New mapping technology is boon to planners. *The Boston Globe* (August 22): 1.

Hirschfield, A., Brown, P. and Todd (1995) GIS and the analysis of spatially-referenced crime data: Experiences in Merseyside, UK. *International Journal of Geographical Information Systems* 9(2): 191–210.

Hook, P. (1997) Crime Patterns. *Police Review* (20 June): 19–20.

Hubbs, R. (1998) The Greenway rapist case: Matching repeat offenders with crime locations. In: N. La Vigne and J. Wartell (eds), *Crime Mapping Case Studies: Successes in the Field*. Washington, DC: Police Executive Research Forum, pp. 93–8.

Hunt, E. D. and E. B. W. Zubrow (1997) Building crime analysis extension for ArcView. *Proceedings, 1997 Environmental Systems Research Institute International User Conference*. Available at http://www.esri.com/library/userconf/archive.html.

Hyatt, R. A. (1999) Measuring crime in the vicinity of public housing with GIS. *Proceedings, 1999 Environmental Systems Research Institute International User Conference*. Available at http://www.esri.com/library/userconf/archive.html.

Hyatt, R. A. and H. R. Holzman (1999) *Guidebook for Measuring Crime in Public Housing with Geographic Information Systems*. Washington, DC: US Department of Housing and Urban Development.

I

Ireland, P. (1998) Helping police with their enquiries. *Mapping Awareness* 12(3): 20–24.

J

Jefferis, E. (1999) *A Multi-Method Exploration of Crime Hot Spots: A Summary of Findings*. Washington, DC: US Department of Justice, National Institute of Justice, Crime Mapping Research Center.

K

Kelly, J. (1999) MapInfo helps take a byte out of crime. *Crime Mapping News* 1(4): 5–7.

Kennedy, D. M., A. A. Braga, and A. M. Piehl (1997) (Un)known universe: Mapping gangs and gang violence in Boston. In: D. Weisburd and T. McEwen (eds) *Crime Mapping and Crime Prevention*. Monsey, NY: Criminal Justice Press, pp. 219–62.

Keppel, R. D. and J. G. Weis (1993) HITS: Catching criminals in the Northwest. *FBI Law Enforcement Bulletin* 62(4): 14–19.

Kinslow, M. W. and T. J. Valentine (1997) Design of ArcView interface for crime analysis. *Proceedings, 1999 Environmental Systems Research Institute International User Conference*. Available at http://www.esri.com/library/userconf/archive.html.

Klein, S. and R. Getz (1999) Blueline CPD: Interactive TV fights crime. *Law and Order* 47(10): 237–42.

Kondo, I. K. and P. C. Unsinger (1972) System of using aerial photographs in police patrol work. *Law and Order* 20(11): 82–5.

L

La Vigne, N. and J. Wartell (eds) (1998) *Crime Mapping Case Studies: Successes in the Field.* Washington, DC: Police Executive Research Forum.

La Vigne, N. and J. Wartell (eds) (2000) *Crime Mapping Case Studies: Successes in the Field, Volume 2.* Washington, DC: Police Executive Research Forum.

Larson, R. (1989) The new crime stoppers. *Technology Review* 92 (Nov./Dec.): 26–31.

LeBeau, J. (1987) The methods and measures of centrography and the spatial dynamics of rape. *Journal of Quantitative Criminology* 3(2): 125–41.

LeBeau, J. L. (1992) Four case studies illustrating the spatial-temporal analysis of serial rapists. *Police Studies* 15(3): 124–45.

LeBeau, J. L. (1995) The temporal ecology of call for police service. In: C. R. Block, M. Dabdoub, and S. Fregly (eds), *Crime Analysis Through Computer Mapping.* Washington, DC: Police Executive Research Forum, pp. 111–28.

LeBeau, J. L. (1997) *Summary of mapping violence and high frequency calls for police service: The Charlotte, North Carolina example.* Washington, DC: National Institute of Justice.

LeBeau, J. L. and K. L. Vincent (1997) Mapping it out: Repeat-address burglary alarms and burglaries. In: D. Weisburd and J. T. McEwen (eds), *Crime Mapping and Crime Prevention.* Monsey, NY: Criminal Justice Press, pp. 289–310.

Lee, Y. (1996) Managing probation programs with geographic information systems. *Computers, Environment, and Urban Systems* 19(5/6): 409–17.

Lee, Y. and F. J. Egan (1972) The geography of urban crime: The spatial pattern of serious crime in the City of Denver. *Proceedings, Association of American Geographers* 4: 59–64.

Leipnik, M. (1999) GIS use by Texas police departments. *Proceedings, 1999 Environmental Systems Research Institute International User Conference.* Available at http://www.esri.com/library/userconf/archive.html.

Leipnik, M., D. P. Albert, D. Kidwell and A. Mellis (2000) How law enforcement agencies can make geographic information technologies work for them. *Police Chief* 67(9): 34–49.

Levine, N. (1996) Spatial statistics and GIS: Software tools to quantify spatial patterns. *Journal of the American Planning Association* 62(3): 381–91.

Levine, N. (1999) *Development of a Spatial Analysis Toolkit for use in a Metropolitan Crime Incident Geographic Information System.* Washington, DC: National Institute of Justice.

Lewin, J. and K. Morrison (1995) Use of mapping to support community-level decision making. In: C. R. Block, M. Dabdoub, and S. Fregly (eds), *Crime Analysis Through Computer Mapping.* Washington, DC: Police Executive Research Forum, pp. 259–276.

Leyden, P. (1993) High-tech van may allow officials to do field work from office. *Star Tribune* (Minneapolis, MN) (May 24): Section B. p. 1.

Lodha, S. K. and A. Verma (1999) Animations of crime maps using virtual reality modeling language. *Western Criminology Review* 1(2): 1–19.

Logan, T. L., A. Bryant, F. W. Mintz, R. W. Muller, D. T. Williams (1995) Mapping marijuana probability grow sites: An operational law enforcement tool. *Proceedings, Counter Law Enforcement: Applied Technology for Improved Operational Effectiveness International Technology Symposium*. Washington, DC: US Executive Office of the President, Part 2, pp. 17-1–17-9.

Lowman, J. (1986) Conceptual issues in the geography of crime: Toward a geography of social control. *Annals of the Association of American Geographers* 76 (Mar): 81–94.

Lu, X. and D. Lawless (1998) Using GIS for crime analysis and mapping. *Proceedings, 1998 Environmental Systems Research Institute International User Conference*. Available at http://www.esri.com/library/userconf/archive.html.

Lutz, W. E. (1996) Computer mapping: A proven tool to fight arson. *Fire and Arson Investigator* 47(1): 15–18.

Lutz, W. E. (1998) Computer mapping helps identify arson targets. *Police Chief* 66(5): 50–2.

Lyew, M. (1996) A new weapon for fighting crime. *GIS Europe* 5(4): xviii–xx.

M

MacKay, R. (1999) Geographic profiling: A new tool for law enforcement. *Police Chief* 66(12): 51–9.

Mallory, J. (1989) Small agencies use computers. *Law and Order* 37(6): 37–41.

Maltz, M. D. (1995) Crime mapping and the drug market analysis program (DMAP). In: C. R. Block, M. Dabdoub, and S. Fregly (eds), *Crime Analysis Through Computer Mapping*. Washington, DC: Police Executive Research Forum, pp. 213–20.

Maltz, M. D., A. C. Gordon, and W. Friedman (1991) *Mapping Crime in its Community Setting: Event Geography Analysis*. New York, NY: Springer Verlag.

Mamalian, C. A., N. G. La Vigne, and the staff of the Crime Mapping Research Center. (1999) *The Use of Computerized Crime Mapping by Law Enforcement: Survey Results*. Washington, DC: US Department of Justice, National Institute of Justice.

Marc-Aurele, J. (1990) Business of computers. *Law Enforcement Technology* 17(5): 52, 54–5.

Marshall, J. (1995) 999 Constabulary duties in Nottinghamshire made easier with GIS. *GIS Europe* 4(5): 44–6.

Martin, D., E. Barnes, and D. Britt (1998) The multiple impacts of mapping it out: Police, geographic information systems (GIS) and community mobilization during Devil's Night in Detroit, Michigan. In: N. La Vigne and J. Wartell (eds), *Crime Mapping Case Studies: Successes in the Field*. Washington, DC: Police Executive Research Forum, pp. 3–14.

Mazerolle, L. G., C. Bellucci, and F. Gajewski (1997) Crime mapping in police departments: The challenges of building a mapping system. In: D. Weisburd and J. T. McEwen (eds), *Crime Mapping and Crime Prevention*. Monsey, NY: Criminal Justice Press, pp. 131–56.

Mazerolle, L. G. and T. E. Conover (1998) A multi-method exploration of crime hot-spots: Spatial and temporal analysis of crime (STAC). Paper presented at the

Annual Meeting of the Academy of Criminal Justice Sciences. Albuquerque, NM, March 11, 1998.

McCabe, C. (1990) From open fields to open skies: The constitutionality of Aerial surveillance. *Narc Officer* 6(4): 47, 49–51.

McEwen, J. T. and F. S. Taxman (1995) Applications of computer mapping to police operations. In: J. E. Eck and D. Weisburd (eds), *Crime and Place*. Monsey, NY: Criminal Justice Press; and Washington, DC: Police Executive Research Forum, pp. 259–84.

McLafferty, S., D. Williamson, and P. G. McGuire (2000) Identifying crime hot spots using kernel smoothing. In: V. Goldsmith, P. G. McGuire, J. H. Mollenkopf, and T. A. Ross (eds), *Analyzing crime patterns: Frontiers of Practice*, pp. 77–86.

McLaughlin, M. (1994) *Automatic Vehicle Location for Law Enforcement*. Sacramento: Search Group, Inc.

McLean, H. E. (1990) Getting high on crime. *Law and Order* 38(7): 30–6.

Meeker, J. W. (1999) Accountability for inappropriate use of crime maps and the sharing of inaccurate data. *Crime Mapping and Data Confidentiality Roundtable July 8–9, 1999*. Washington, DC. Crime Mapping Research Center, National Institute of Justice.

Miller, T. (1993) GIS Catches Criminals. *GIS World* 6(5): 42–3.

Miller, T. (1993) GIS helps San Bernardino Police take a byte out of crime. *American City and Country* 108(3): 42.

Miller, T. (1994) Computers track the criminals' trail. *American Demographics* 16 (Jan.): 13–14.

Miller, T. (1995) Integrating crime mapping with CAD and RMS. In: C. R. Block, M. Dabdoub, and S. Fregly (eds), *Crime Analysis Through Computer Mapping*. Washington, DC: Police Executive Research Forum, pp. 179–88.

Mitchell, D. (1997) *Crime, Policing and GIS: An emerging technology?* London: Association for Geographic Information, Publication No. 5. Available at http://www.agi.org.uk/pages/freepubs/crime.html.

Moland, R. S. (1998) Graphical display of murder trial evidence. In: N. La Vigne and J. Wartell (eds), *Crime Mapping Case Studies: Successes in the Field*. Washington, DC: Police Executive Research Forum, pp. 69–79.

Monmonier, M. (1997) *Cartographies of Danger: Mapping Hazards in America*. Chicago, IL: University of Chicago Press.

Moore, M. R. (1995) Keeping it simple. In: C. R. Block, M. Dabdoub, and S. Fregly (eds), *Crime Analysis Through Computer Mapping*. Washington, DC: Police Executive Research Forum, pp. 161–6.

Morrison, R. D. (1998) In jail at home: The Pro Tech Monitoring inc. SMART monitoring system handles house arrest monitoring via satellite. *Law Enforcement Technology* 25(6): 86–9.

Murphy, J. H. (1995) Tracking in urban environments: Alternative technological approaches. *Proceedings, Counter Law Enforcement: Applied Technology for Improved Operational Effectiveness International Technology Symposium*. Washington, DC: US Executive Office of the President, Part 2, 17-1–17-9.

N

Nelson, L. (1999) Crime mapping and Environmental Systems Research Institute. *Crime Mapping News* 1(4): 1–8.

Nowicki, D. E. (1999) A "proposal" for how partnerships between police agencies and researchers on geocoded data sharing. *Crime Mapping and Data Confidentiality Roundtable July 8–9, 1999*. Washington, DC. Crime Mapping Research Center, National Institute of Justice.

Nuttall, N. (1991) Putting the criminal on the map. *The Times* (August 29).

O

O'Kane, J., R. Fisher, and L. Green (1994) Mapping campus crime. *Security Journal* 5(3): 172–80.

Olligschlaeger, A. M. (1997) Artificial neural networks and crime mapping. In: D. Weisburd and J. T. McEwen (eds), *Crime Mapping and Crime Prevention*. Monsey, NY: Criminal Justice Press, pp. 313–47.

Openshaw, S. (1993) GIS crime and GIS criminality. *Environment and Planning A* 25(4): 451–58.

Openshaw, S., A. Cross, M. Charlton, and C. Brunsdon (1990) Lessons learnt from a post mortem of a failed GIS. *2nd National Conference and Exhibition of the AGI*, Brighton.

P

Page, J. (1997) Dial M for mapping. *Mapping Awareness* 11(2): 25–7.

Panel Discussion (1983) Using computers in crime fighting, training, and administration: A practical approach. *The Police Chief* 50 (Mar.): 123–30.

Petronis, K. R., C. C. Johnson, and E. D. Wish (1995) *Location of Drug-Using Arrestees and Treatment Centers in Washington, D.C.: A Geocoding Demonstration Project*. College Park, M.D.: Center for Substance Abuse Research, University of Maryland.

Phelan, L. and J. Fenske (1995) Crime analysis: Administrative aspects. *TELEMASP Bulletin* 1(10).

Phillips, P. D. (1972) A prologue to the geography of crime. *Proceedings, Association of American Geographers*, 4: 86–91.

Pilant, L. (1995) Automated vehicle location. *Police Chief* 62(9): 37–46.

Pilant, L. (1996) High-technology solutions. *Police Chief* 63(5): 38–51.

Pilant, L. (1997) Spotlight on computerized crime mapping. *Police Chief* 64(12): 60–1, 63–9.

Pilant, L. (1999) Crime mapping and analysis. *Police Chief* 66 (12): 38–47.

Pilant, L. (1999) Spotlight on ... crime mapping and analysis. *The Police Chief* 66(12): 38–47.

Pollitt, M. (1994) Protecting Irish interests: GIS on patrol. *GIS World* 3(11): 18–20.

Pyle, G. F. (1974) *The Spatial Dynamics of Crime*. Chicago: Department of Geography, University of Chicago.

Pyle, G. F. (1976) Spatial and temporal aspects of crime in Cleveland, Ohio. *American Behavioral Scientist* 20(2): 175–97.

Q

Quist, J. (1999) GIS crime mapping improves public safety. *Nation's Cities Weekly* 22(17): 7, 10.

R

Ratcliffe, J. H. (2000) Implementing and integrating crime mapping into a police intelligence environment. *International Journal of Police Science Management* 2(4): 313–23.

Ratcliffe, J. H. and M. J. McCullagh (1998) Aoristic crime analysis. *International Journal of Geographical Information Science* 12(7): 751–64.

Ratcliffe, J. H. and M. J. McCullagh (1998) Identifying repeat victimization with GIS. *British Journal of Criminology* 38(4): 651–80.

Ratcliffe, J. H. and M. J. McCullagh (1998) The perception of crime hot spots: A spatial study in Nottingham, U.K. In: N. La Vigne and J. Wartell (eds), *Crime Mapping Case Studies: Successes in the Field*. Washington, DC: Police Executive Research Forum, pp. 45–51.

Read, T. and D. Oldfield (1995) Local crime analysis. London: Police Research Group; Crime Detection and Prevention Series, Paper No. 65.

Reboussin, R., J. Warren, and R. R. Hazelwood (1995) Mapless mapping in analyzing the spatial distribution of serial rapes. In: C. R. Block, M. Dabdoub, and S. Fregly (eds), *Crime Analysis Through Computer Mapping*. Washington, DC: Police Executive Research Forum, pp. 59–64.

Regional Community Policing Training Institute. Geographic information systems. RCPTI Online. Wichita, KS. Regional Community Policing Training Institute. Available at http://www.wsurcpi.org/tech/gis.html.

Rengert, G. F. (1975) Some effects of being female on criminal spatial behavior. *The Pennsylvania Geographer* 13: 10–18.

Rengert, G. F. (1995) Comparing cognitive hot spots to crime hot spots. In: C. R. Block, M. Dabdoub, and S. Fregly (eds), *Crime Analysis Through Computer Mapping*. Washington, DC: Police Executive Research Forum, pp. 33–48.

Rengert, G. F. (1998) *The Geography of Illegal Drugs*. Boulder, CO: Westview Press.

Rengert, G. F. and J. Wasilchick (1985) *Suburban Burglary: A Time and Place for Everything*. Springfield, IL: Charles C. Thomas.

Rengert, G. F. and W. V. Pelfrey (1997) Cognitive mapping of the city center: Comparative perceptions of dangerous places. In: D. Weisburd and J. T. McEwen (eds), *Crime Mapping and Crime Prevention*. Monsey, NY: Criminal Justice Press, pp. 193–218.

Rengert, G. F., M. Mattson, and K. Henderson (1998) The development and use of high definition geographic information systems. Paper presented to the second annual Crime Mapping Research Conference, Arlington, VA, December 11.

Reppetto, T. A. (1974) *Residential Crime*. Cambridge, MA: Ballinger.

Reuland, M. M. (ed.) (1997) *Information Management and crime Analysis: Practitioners' Recipes for Success*. Police Executive Research Forum.

Rewers, R. F. and L. A. Green (1995) The Chicago area four GeoArchive: An information foundation. In: C. R. Block, M. Dabdoub, and S. Fregly (eds), *Crime Analysis Through Computer Mapping*. Washington, DC: Police Executive Research Forum, pp. 221–9.

Rex, B. and R. Rasmussen (2000) Starlight and map objects for data mining crime information. *Proceedings, 2000 Environmental Systems Research Institute International User Conference*. Available at http://www.esri.com/library/userconf/archive.html.

Rich, T. F. (1995) *The Use of Computerized Mapping in Crime Control and Prevention Programs. Research in Action* (July). Washington, DC: US Department of Justice, National Institute of Justice.

Rich, T. F. (1996) *The Chicago Police Department's Information Collection for Automated Mapping (ICAM) Program: A Program Focus*. Washington, DC: US National Institute of Justice.

Rich, T. F. (1999) Mapping the path to problem solving. *National Institute of Justice Journal* (Oct.): 2–9.

Robey, R. (1998) Governments embrace new mapping system. *The Denver Post*, 2nd edn (July 6): Section B, p. 2.

Rogers, D. (1999) Getting crime analysis on the map. *Law Enforcement Technology* 26(11): 76–9.

Rogers, R. and D. Craig (1994) Geographic information systems in policing. *Police Studies* 17(2): 67–78.

Rogers, R. and D. Craig (1996) Geographic information systems: Computers in law enforcement. *Journal of Crime and Justice* 19(1): 61–74.

Roncek, D. and P. Maier (1991) Bars, blocks, and crimes revisited: Linking the theory of routine activities to the empiricism of hot spots. *Criminology* 29(4): 725–53.

Roncek, D. and A. Montgomery (1995) Spatial autocorrelation revisited: Conceptual underpinnings and practical guidelines for the use of the generalized potential as a remedy for spatial autocorrelation in large samples. In: C. R. Block, M. Dabdoub, and S. Fregly (eds), *Crime Analysis Through Computer Mapping*. Washington, DC: Police Executive Research Forum, pp. 99–110.

Rose, S. (1998) Mapping technology takes a bite out of crime. *Business Geographics* (June): p. 98.

Rosenbaum, D. P. and P. J. Lavrakas (1995) Self-reports about place: The application of survey and interview methods to the study of small areas. In: D. Weisburd and T. McEwen (eds). *Crime Prevention Studies*, vol. 4. Monsey, NY: Criminal Justice Press, pp. 285–313.

Rossmo, D. K. (1995) Overview: Multivariate spatial profiles as a tool in crime investigation. In: C. R. Block, M. Dabdoub, and S. Fregly (eds), *Crime Analysis Through Computer Mapping*. Washington, DC: Police Executive Research Forum, pp. 65–97.

Rossmo, D. K. (1995) Strategic crime patterning: Problem-oriented policing and displacement. In: C. R. Block, M. Dabdoub, and S. Fregly (eds), *Crime Analysis Through Computer Mapping*. Washington, DC: Police Executive Research Forum, pp. 1–14.

S

Sanford, R. (1995) How to develop a tactical early warning system on a small-city budget. In: C. R. Block, M. Dabdoub, and S. Fregly (eds), *Crime Analysis Through Computer Mapping*. Washington, DC: Police Executive Research Forum, pp. 199–209.

Sherman, L. W. (1995) Hot spots of crime and criminal careers of places. In: J. E. Eck and D. Weisburd (eds), *Crime and Place*. Monsey, NY: Criminal Justice Press; and Washington, DC: Police Executive Research Forum, pp. 35–52.

Sherman, L. W., P. R. Gartin, and M. E. Buergan (1989) Hot spots of predatory crime: Routine activities and the criminology of place. *Criminology* 27(1): 27–56.

Sherman, L. W., D. C. Gottfredson, D. L. MacKenzie, J. Eck, P. Reuter, and S. D. Bushway (1998) *Preventing Crime: What Works, What Doesn't, What's Promising*. Washington, DC: US Department of Justice, National Institute of Justice.

Shields, P. J. (1990) Wichita Police Department Air Section: Law enforcement from above. *National F.O.P. Journal* 19(2): 26–8.

Showley, R. M. (1998) Crime busters get a break in their investigations. *The San Diego Union-Tribune* (July 26): Section H, p. 1.

Showley, R. M. (1998) Technology growing by leaps and boundaries. *The San Diego Union-Tribune* (July 26): Section A, p. 1.

Sims, D. (1993) Desktop applications support your local sheriff. *IEEE Computer Graphics and Applications* 13 (July): 14.

Siuru, B. (1999) Tracking 'down': Space-age GPS technology is here. *Corrections Technology and Management* 3(5): 12–14.

Smith, C. and G. Patterson (1980) Cognitive mapping and the subjective geography of crime. In: D. Georges-Abeyie and K. Harries (eds), *Crime: A Spatial Perspective*. New York, NY: Columbia University Press.

Smith, S. (1986) *Crime, Space, and Society*. Cambridge: Cambridge University Press.

Sorensen, S. L. (1997) SMART mapping for law enforcement settings: Integrating GIS and GPS for dynamic, near-real time applications and analysis. In: D. Weisburd and J. T. McEwen (eds), *Crime Mapping and Crime Prevention*. Monsey, NY: Criminal Justice Press, pp. 349–78.

Sparrow, M. (1991) The application of network analysis to criminal intelligence: An assessment of the prospects. *Social Networks* 13: 252–74.

Stahura, J. M. and C. R. Huff (1979) The new 'zones of transition': gradients of crime in metropolitan areas. *Review of Public Data Use* 7: 41–8.

Stallo, M. (1995) Mapping software and its values to law enforcement. In: C. R. Block, M. Dabdoub, and S. Fregly (eds), *Crime Analysis Through Computer Mapping*. Washington, DC: Police Executive Research Forum, pp. 229–34.

Stephens, M. (1995) Commission and auditor in a map flap. *The Columbus Dispatch* (August 14): Section C, p. 1.

Strandberg, K. W. (1998) Pursuit at high speeds. *Law Enforcement Technology* 25(9): 50–4.

T

Taxman, F. S. and T. McEwen (1997) Using geographical tools with interagency work groups to develop and implement control strategies. In: D. Weisburd and T. McEwen (eds), *Crime Mapping and Crime Prevention*. Monsey, NY: Criminal Justice Press, pp. 83–111.

Temple, H. R., and W. L. Stewart (1990) National Guard in the war on drugs. *Military Review* (March): 41–8.

Trickett, A., D. Ellingworth, T. Hope, and K. Pease (1995) Crime victimization in the eighties – changes in area and regional inequality. *British Journal of Criminology* 35(3): 343–59.

Tyler, S. (1990) Computer assistance for the California earthquake rescue effort. *Police Chief* 57(3): 42–3.

U

US Department of Justice (1999) Mapping out crime: Providing 21st century tool for safe communities. Report on the task force on crime mapping and data-driven management. Washington, DC: US Department of Justice.

V

Van Zandt, C. R. (1994) Real silence of the lambs. *Police Chief* 61(4): 45, 47–50, 52.

Vogel, M. (1994) Geographic information system will be planning tool of future. *The Buffalo News* (August 9): 8.

W

Wasserman, I. M. and S. Stack (1995) Graphic spatial autocorrelation and United States suicide patterns. *Archives of Suicide Research* 1(2): 121–9.

Wasserman, I. M. and S. Stack (1995) Spatial autocorrelation patterns with regard to suicide in the United States. In: C. R. Block, M. Dabdoub, and S. Fregly (eds), *Crime Analysis Through Computer Mapping*. Washington, DC: Police Executive Research Forum, pp. 49–54.

Webster, B. and E. F. Connors (1993) Police methods for identifying community problems. *American Journal of Police* 12(1): 75–102.

Weisburd, D. and L. Green (1995) Measuring immediate spatial displacement: Methodological issues and problems. In: J. E. Eck and D. Weisburd (eds), *Crime and Place*. Monsey, NY: Criminal Justice Press; and Washington, DC: Police Executive Research Forum, pp. 349–61.

Weisburd, D. and L. Green (1995) Policing drug hot spots: The Jersey City DMA (drug market analysis) experiment. *Justice Quarterly* 12(4): 711–41.

Weisburd, D. and J. T. McEwen (eds) (1997) *Crime Mapping and Crime Prevention*. Monsey, NY: Criminal Justice Press.

Wendelken, S. (1995) GIS enhances preventative law enforcement. *GIS World* 8(1): 58–61.

Westerfeld, F. E. (1999) Case study of half-way house location. Crime Mapping Research Center listserv, listproc@aspensys.com, retrieved on January 22.

Wilkinson, R. A. and P. Ritchie-Matsumoto (1997) Collaboration and applications: Exchange of information between facilities can improve safety, cut costs. *Corrections Today* 59(4): 64, 66–7.

Williams, A. K. (1983) Aerial surveillance to detect growing marihuana. *FBI Law Enforcement Bulletin* 52(2): 9–14.

Williams, R. (1986) Computerized crime map spots crime patterns. *Law and Order* 34(1): 28–31.

Williamson, D. and V. Goldsmith (1997) *Evaluation of Software for Displaying and Analyzing Crime Patterns and Trends: Results from Five GIS and Spatial Statistical Software Analyses of Murder from the Bronx, New York*. Washington. DC: National Institute of Justice.

Wise, B. (1995) Catching crooks with computers. *American City and Country* 110 (May): 54.

Witkin, G. (1997) Making mean streets nice. Computer maps that take the 'random' out of violence. *US News & World Report* 121(26): 63.

Wood, D. R. (1998) Geospatial analysis of rural burglaries. In: N. La Vigne and J. Wartell (eds), *Crime Mapping Case Studies: Successes in the Field*. Washington, DC: Police Executive Research Forum, pp. 117–21.

Index